Lecture Notes in Mathematics   2243

More information about this subseries at http://www.springer.com/series/7098

Saint-Flour Probability Summer School

The Saint-Flour volumes are reflections of the courses given at the Saint-Flour Probability Summer School. Founded in 1971, this school is organised every year by the Laboratoire de Mathématiques (CNRS and Université Clermont Auvergne, Clermont-Ferrand, France). It is intended for PhD students, teachers and researchers who are interested in probability theory, statistics, and in their applications.

The duration of each school is 12 days (it was 17 days up to 2005), and up to 100 participants can attend it. The aim is to provide, in three high-level courses, a comprehensive study of some fields in probability theory or Statistics. The lecturers are chosen by an international scientific board. The participants themselves also have the opportunity to give short lectures about their research work.

Participants are lodged and work in the same building, a former seminary built in the 18th century in the city of Saint-Flour, at an altitude of 900 m. The pleasant surroundings facilitate scientific discussion and exchange.

The Saint-Flour Probability Summer School is supported by:
- Laboratoire de Mathématiques Blaise Pascal
- Université Clermont Auvergne
- Centre National de la Recherche Scientifique (C.N.R.S.)

For more information, see

http://recherche.math.univ-bpclermont.fr/stflour/stflour-en.php

Christophe Bahadoran
bahadora@uca.fr

Arnaud Guillin
Arnaud@uca.fr

Hacène Djellout
Hacène.Djellout@uca.fr

stflour@math.univ-bpclermont.fr

Université Clermont Auvergne - Aubière cedex, France

Asaf Nachmias

Planar Maps, Random Walks and Circle Packing

École d'Été de Probabilités de Saint-Flour
XLVIII - 2018

Asaf Nachmias
Department of Mathematical Sciences
Tel Aviv University
Tel Aviv, Israel

ISSN 0075-8434   ISSN 1617-9692   (electronic)
Lecture Notes in Mathematics
ISSN 0721-5363
École d'Été de Probabilités de Saint-Flour
ISBN 978-3-030-27967-7   ISBN 978-3-030-27968-4   (eBook)
https://doi.org/10.1007/978-3-030-27968-4

Mathematics Subject Classification (2010): Primary: 82B41; Secondary: 52C26

This book is an open access publication.

This Springer imprint is published by the registered company Springer Nature Switzerland AG.
The registered company address is: Gewerbestrasse 11, 6330 Cham, Switzerland

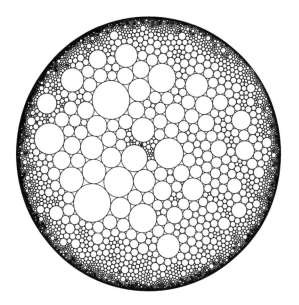

Preface

These lecture notes are intended to accompany a single-semester graduate course. They are meant to be *entirely self-contained*. All the theory required to prove the main results is presented and only basic knowledge in probability theory is assumed.

In Chap. 1, we describe the main storyline of this text. It is meant to be light bedtime reading exposing the reader to the main results that will be presented and providing some background. Chapter 2 introduces the theory of electric networks and discusses their highly useful relations to random walks. It is roughly based on Chap. 8 of Yuval Peres' excellent lecture notes [69]. We then discuss the circle packing theorem and present its proof in Chap. 3. Chapter 4 discusses the beautiful theorem of He and Schramm [40], relating the circle packing type of a graph to recurrence and transience of the random walk on it. To the best of our knowledge, their work is the first to form connections between the circle packing theorem and probability theory. Next in Chap. 5, we present the highly influential theorem of Benjamini and Schramm [11] about the almost sure recurrence of the simple random walk in planar graph limits of bounded degrees. The notion of a *local limit* (also known as *distributional limit* or *Benjamini-Schramm limit*) of a sequence of finite graphs was introduced there for the first time to our knowledge (and also studied by Aldous–Steele [3] and Aldous–Lyons [2]); this notion is highly important in probability theory as well as other mathematical disciplines (see [2] and the references within). In Chap. 6, we provide a theorem from which one can deduce the almost sure recurrence of the simple random walk on many models of random planar maps. This theorem was obtained by Ori Gurel-Gurevich and Nachmias in [31]. Chapter 7 discusses uniform spanning forests on planar maps and appeals to the circle packing theorem to show that the free uniform spanning forest on proper planar maps is almost surely connected, i.e., it is in fact a tree. This theorem was obtained by Hutchcroft and Nachmias in [45]. We close these notes in Chap. 8 with a description of some related contemporary developments in this field that are not presented in this text.

We have made an effort to add value beyond what is in the published papers. Our proof of the circle packing theorem in Chap. 3 is inspired by Thurston's argument [82] and Brightwell–Scheinerman [13], but we have made what we think are some simplifications; the proof also employs a neat argument due to Ohad Feldheim and Ori Gurel-Gurevich (Theorem 3.14) which makes the drawing part of the argument rather straightforward and avoids topological considerations that are used in the classical proofs. The original proof of the He–Schramm Theorem [40] is based on the notion of *discrete extremal length* which is essentially a form of *effective resistance* in electric networks (in fact, the *edge* extremal length is precisely an effective resistance, see [61, Exercise 2.78]). We find that our approach in Chap. 4 using electric networks is somewhat more robust and intuitive to probabilists. We obtain a quantitative version of the He-Schramm Theorem in Chap. 4 as well as the Benjamini–Schramm Theorem [11] in Chap. 5 (see Theorem 5.8). These quantified versions are key to the proofs of Chap. 6. Lastly, some aspects of stationary random graphs are better explained here in Chap. 6 than in the publication [31]. The videoed lectures of this course taken in 48th Saint-Flour summer school are available at the author's webpage http://www.math.tau.ac.il/~asafnach/.

Acknowledgments

I would like to deeply thank Daniel Jerison, Peleg Michaeli, and Matan Shalev for typing, editing, and proofreading most of this text and for the many comments, corrections, and suggestions. I am indebted to Tom Hutchcroft for his assistance in writing the introduction and for surveying related topics not included in these notes (Chaps. 1 and 8). I thank Sébastien Martineau, Pierre Petit, Dominik Schmid, and Mateo Wirth for their corrections and comments to this text. I am also grateful to the participants of the 48th Saint-Flour Summer School and its organizers, Christophe Bahadoran, Arnaud Guillin, and Hacène Djellout, for a very enjoyable summer school.

I am beholden to my collaborators on these topics: Omer Angel, Martin Barlow, Itai Benjamini, Nicolas Curien, Ori Gurel-Gurevich, Tom Hutchcroft, Daniel Jerison, Gourab Ray, Steffen Rohde, and Juan Souto. I have learned a lot from our work and conversations. Special thanks go to Ori Gurel-Gurevich for embarking together on this research endeavor beginning in 2011 at the University of British Columbia, Vancouver, Canada. Many of the ideas and methods presented in these notes were obtained in our joint work.

I am also highly indebted to the late Oded Schramm whose mathematical work, originality, and vision, especially in the topics studied in these notes, have been an enormous source of inspiration. It is no coincidence that his name appears on almost every other page here. It has become routine for my collaborators and I to ask ourselves "What would Oded do?" hoping that reflecting on this question would cut right to the heart of matters. Steffen Rohde's wonderful survey [71] of Oded's work is very much recommended.

Lastly, I thank Shira Wilkof and baby Ada for muse and inspiration.

Tel Aviv, Israel   Asaf Nachmias[1]
July 2019

[1]This project has received funding from the European Research Council (ERC) under the European Union's, Horizon 2020 research and innovation programme (grant agreement number 676970, RANDGEOM).

Contents

Chapter 1
Introduction

1.1 The Circle Packing Theorem

A planar graph is a graph that can be drawn in the plane, with vertices represented by points and edges represented by non-crossing curves. There are many different ways of drawing any given planar graph and it is not clear what is a canonical method. One very useful and widely applicable method of drawing a planar graph is given by Koebe's 1936 *circle packing theorem* [51], stated below. As we will see, various geometric properties of the circle packing drawing (such as existence of accumulation points and their structure, bounds on the radii of circles and so on) encode important probabilistic information (such as the recurrence/transience of the simple random walk, connectivity of the uniform spanning forest and much more). This deep connection is especially fruitful to the study of random planar maps. Indeed, one of the main goals of these notes is to present a self-contained proof that the so-called *uniform infinite planar triangulation* (UIPT) is almost surely recurrent [31].

A **circle packing** is a collection of discs $P = \{C_v\}$ in the plane $\mathbb{C}$ such that any two distinct discs in P have disjoint interiors. That is, distinct discs in P may be tangent, but may not overlap. Given a circle packing P, we define the **tangency graph** $G(P)$ of P to be the graph with vertex set P and with two vertices connected by an edge if and only if their corresponding circles are tangent. The tangency graph $G(P)$ can be drawn in the plane by drawing straight lines between the centers of tangent circles in P, and is therefore planar. It is also clear from the definition that $G(P)$ is **simple**, that is, any two vertices are connected by at most one edge and there are no edges beginning and ending at the same vertex. See Fig. 1.1.

We call a circle packing P a circle packing of a planar graph G if $G(P)$ is isomorphic to G.

Theorem 1.1 (Koebe 1936) *Every finite simple planar graph G has a circle packing. That is, there exists a circle packing P such that $G(P)$ is isomorphic to G.*

© The Author(s) 2020
A. Nachmias, *Planar Maps, Random Walks and Circle Packing*, Lecture Notes
in Mathematics 2243, https://doi.org/10.1007/978-3-030-27968-4_1

Fig. 1.1 A planar graph and a circle packing of it

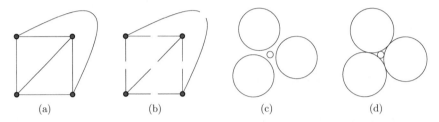

Fig. 1.2 A sketch of how to obtain circle packings using Koebe's extension of the Riemann mapping theorem to finitely connected domains, which states that every domain $D \subseteq \mathbb{C} \cup \{\infty\}$ with at most finitely many boundary components is conformally equivalent to a *circle domain*, that is, a domain all of whose boundary components are circles or points. (**a**) Step 1: We begin by drawing the finite simple planar graph G in the plane in an arbitrary way. (**b**) Step 2: If we remove the 'middle ε' of each edge, then the complement of the resulting drawing is a domain with finitely many boundary components. (**c**) Step 3: Finding a conformal map from this domain to a circle domain gives an 'approximate circle packing' of G. (**d**) Step 4: Taking the limit as $\varepsilon \downarrow 0$ can be proven to yield a circle packing of G

One immediate consequence of the circle packing theorem is Fáry's Theorem [25], which states that every finite simple planar graph can be drawn so that all the edges are represented by straight lines.

The circle packing theorem was first discovered by Koebe [51], who established it as a corollary to his work on the generalization of the Riemann mapping theorem to finitely connected domains; a brief sketch of Koebe's argument is given in Fig. 1.2. The theorem was rediscovered and popularized in the 1970s by Thurston [82], who showed that it follows as a corollary to the work of Andreev on hyperbolic polyhedra (see also [63]). Thurston also initiated a popular program of understanding circle packing as a form of *discrete complex analysis*, a viewpoint which has been highly influential in the subsequent development of the subject and which we discuss in more detail below (see [79] for a review of a different form of discrete complex analysis with many applications to probability). There are now many proofs of the circle packing theorem available including, remarkably, four distinct proofs discovered by Oded Schramm. In Chap. 3 we will give an entirely combinatorial proof, which is adapted from the proof of Thurston [63, 82] and Brightwell and Scheinerman [13].

Uniqueness

We cannot expect a uniqueness statement in Theorem 1.1 (see Fig. 1.1; we may "slide" circles 5 and 6 along circle 2). However, when our graph is a *finite triangulation*, circle packings enjoy uniqueness up to circle-preserving transformations.

Definition 1.2 A planar **triangulation** is a planar graph that can be drawn so that every face is incident to exactly three edges. In particular, when the graph is finite this property must hold for the outer face as well.

Claim 1.3 If G is a finite triangulation, then the circle packing whose tangency graph is isomorphic to G is unique, up to Möbius transformations and reflections in lines.

The uniqueness of circle packing was first proven by Thurston, who noted that it follows as a corollary to Mostow's rigidity theorem. Since then, many different proofs have been found. In Chap. 3 we will give a very short and elementary proof of uniqueness due to Oded Schramm that is based on the maximum principle.

Infinite Planar Graphs

So far, we have only discussed the existence and uniqueness of circle packings of *finite* planar triangulations. What happens with infinite triangulations? To address this question, we will need to introduce some more definitions.

Definition 1.4 We say that a graph G is **one-ended** if the removal of any finite set of vertices leaves at most one infinite connected component.

Definition 1.5 Let $P = \{C_v\}$ be a circle packing of a triangulation. We define the **carrier** of P to be the union of the closed discs bounded by the circles of P together with the spaces bounded between any three circles that form a face (i.e., the interstices). We say that P is **in** D if its carrier is D.

See Fig. 1.3 for examples where the carrier is a disc or a square. The circle packing of the standard triangular lattice (see Fig. 4.2) has the whole plane $\mathbb{C}$ as its carrier. It is not too hard to see that if $G(P)$ is an infinite triangulation, then it is one-ended if and only if the carrier of P is simply connected, see Lemma 4.1.

It can be shown via a compactness argument that any simple infinite planar triangulation can be circle packed in *some* domain. Indeed, one can simply take subsequential limits of circle packings of finite subgraphs (the fact that such subsequential limits can be taken is a consequence of the so-called Ring Lemma, see Lemma 4.2). This is performed in Claim 4.3. However, this compactness argument does not give us any control of the domain we end up with as the carrier of our circle packing. The following theorems of He and Schramm [39, 40] give us much

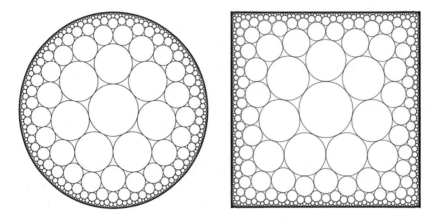

Fig. 1.3 The 7-regular hyperbolic tessellation circle packed in a disc and in a square

better control; they can be thought of as discrete analogues of the Poincaré-Koebe uniformization theorem for Riemann surfaces.

Theorem 1.6 (He and Schramm 1993) *Any one-ended infinite triangulation can be circle packed such that the carrier is either the plane or the open unit disk, but not both.*

This theorem will be proved in Chap. 4 (with the added assumption of finite maximal degree). The proofs in [39, 40] are based on the notion of *discrete extremal length*. We will present our own approach to the proof in Chap. 4 based on a very similar notion of electric resistance discussed in Chap. 2. This approach is somewhat more appealing to a probabilist and allows for quantitative versions of the He-Schramm Theorem that will be used later for the study of random planar maps in Chap. 6.

In view of Theorem 1.6, we call an infinite one-ended simple planar triangulation **CP parabolic** if it can be circle packed in $\mathbb{C}$, and call it **CP hyperbolic** if it can be circle packed in the open unit disk $\mathbb{U}$.

Theorem 1.7 (He and Schramm 1995) *Let T be a CP hyperbolic infinite one-ended simple planar triangulation and let $D \subsetneq \mathbb{C}$ be a simply connected domain. Then there exists a circle packing of T with carrier D.*

What about uniqueness? Theorem 1.7 shows that, in general, we have much more flexibility when choosing a circle packing of an infinite planar triangulation than we have in the finite case, see Fig. 1.3 again. Indeed, it implies that the circle packing of a CP hyperbolic triangulation is *not* determined up to Möbius transformations and reflections, since, for example, we can circle pack the same triangulation in both the unit disc and the unit square, and these two packings are clearly not related by a Möbius transformation. Fortunately, the following theorem of Schramm [73] shows that we recover Möbius rigidity if we restrict the packing to be in $\mathbb{C}$ or $\mathbb{U}$.

Theorem 1.8 (Schramm 1991) *Let T be a one-ended infinite planar triangulation.*

- *If T is CP parabolic, then its circle packing in $\mathbb{C}$ is unique up to dilations, rotations, translations and reflections.*
- *If T is CP hyperbolic, then its circle packing in $\mathbb{U}$ is unique up to Möbius transformations or reflections fixing $\mathbb{U}$.*

Relation to Conformal Mapping

A central motivation behind Thurston's popularization of circle packing was its role as a discrete analogue of conformal mapping. The resulting theory is somewhat tangential to the main thrust of these notes, but is worth reviewing for its beauty, and for the intuition it gives about circle packing. A more detailed treatment of this and related topics is given in [81].

Recall that a map $\phi : D \to D'$ between two domains $D, D' \subseteq \mathbb{C}$ is conformal if and only if it is holomorphic and one-to-one. Intuitively, we can think of the latter condition as saying that ϕ maps infinitesimal circles to infinitesimal circles. Thus, it is natural to wonder, as Thurston did, whether conformal maps can be approximated by graph isomorphisms between circle packings of the corresponding domains, which *literally* map circles to circles.

For each $\varepsilon > 0$, let $\mathbb{T}_\varepsilon = \{\varepsilon n + \varepsilon \frac{1+\sqrt{3}i}{2}m : n, m \in \mathbb{Z}\} \subseteq \mathbb{C}$ be the triangular lattice with lattice spacing ε, which we make into a simple planar triangulation by connecting two vertices if and only if they have distance ε from each other. This triangulation is naturally circle packed in the plane by placing a disc of radius ε around each point of $\mathbb{T}_\varepsilon$: this is known as the **hexagonal packing**. Now, let D be a simply connected domain, and take z_0 to be a marked point in the interior of D. For each $\varepsilon > 0$ let u_ε be an element of $\mathbb{T}_\varepsilon$ of minimal distance to z_0, and let $v_\varepsilon = u_\varepsilon + \varepsilon$ and $w_\varepsilon = u_\varepsilon + (1 + \sqrt{3}i)\varepsilon/2$. For each $\varepsilon > 0$, let $T_\varepsilon(D)$ be the subgraph of $\mathbb{T}_\varepsilon$ induced by the vertices of distance at least 2ε from ∂D (i.e., the subgraph containing all such vertices and all the edges between them), and let $T'_\varepsilon(D)$ be the component of $T_\varepsilon(D)$ containing u_ε. Finally, let $T''_\varepsilon(D)$ be the triangulation obtained from $T'_\varepsilon(D)$ by placing a single additional vertex ∂_ε in the outer face of $T'_\varepsilon(D)$ and connecting this vertex to every vertex in the outer boundary of $T'_\varepsilon(D)$.

Applying the circle packing theorem to $T''_\varepsilon(D)$ and then applying a Möbius transformation or a reflection if necessary, we obtain a circle packing P_ε of $T''_\varepsilon(D)$ with the following properties:

- The boundary vertex ∂_ε is represented by the unit circle,
- the vertex u_ε is represented by a circle centered at the origin,
- the vertex v_ε is represented by a circle centered on the real line, and
- the vertex w_ε is represented by a circle centered in the upper half-plane.

The function sending each vertex of $T'_\varepsilon(D)$ to the center of the circle representing it in P_ε can be extended piecewise on each triangle by an affine extension. Call the resulting function ϕ_ε.

The following theorem was conjectured by Thurston and proven by Rodin and Sullivan [70].

Theorem 1.9 (Rodin and Sullivan 1987) *Let ϕ be the unique conformal map from D to $\mathbb{U}$ with $\phi(z_0) = 0$ and $\phi'(z_0) > 0$. Then ϕ_ε converge to ϕ as $\varepsilon \downarrow 0$, uniformly on compact subsets of D.*

See Fig. 1.4. The key to the proof of Theorem 1.9 was to establish that the hexagonal packing is the only circle packing of the triangular lattice, which is now a special case of Theorem 1.8.

Various strengthenings and generalizations of Theorem 1.9 have been established in the works [21, 36, 38, 41, 42, 80].

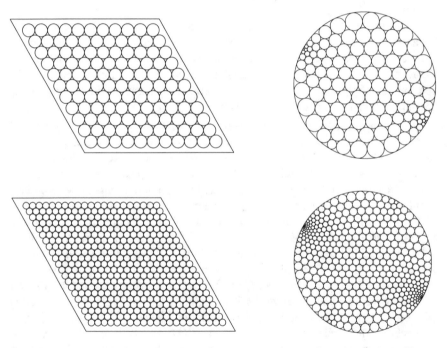

Fig. 1.4 Approximating the conformal map from a rhombus to the disc using circle packing, at two different degrees of accuracy

1.2   Probabilistic Applications

Why should we be interested in circle packing as probabilists? At a very heuristic level, when we uniformize the *geometry* of a triangulation by applying the circle packing theorem, we also uniformize the *random walk* on the triangulation, allowing us to compare it to a standard reference process that we understand very well, namely Brownian motion. Indeed, since Brownian motion is conformally invariant and circle packings satisfy an approximate version of conformality, it is not unreasonable to expect that the random walk on a circle packed triangulation will behave similarly to Brownian motion. This intuition turns out to be broadly correct, at least when the triangulation has bounded degrees, although it is more accurate to say that the random walk behaves like a *quasi-conformal image* of Brownian motion, that is, the image of Brownian motion under a function that distorts angles by a bounded amount.

Although it is possible to make the discussion in the paragraph above precise, in these notes we will be interested primarily in much coarser information that can be extracted from circle packings, namely *effective resistance estimates* for planar graphs. This fundamental topic is thoroughly discussed in Chap. 2. One of the many definitions of the effective resistance $\mathcal{R}_{\mathrm{eff}}(A \leftrightarrow B)$ between two disjoint sets A and B in a finite graph is

$$\frac{1}{\mathcal{R}_{\mathrm{eff}}(A \leftrightarrow B)} = \sum_{v \in A} \deg(v) \mathbb{P}_v(\tau_B < \tau_A^+),$$

where $\mathbb{P}_v$ is the law of the simple random walk started at v, τ_B is the first time the walk hits B, and τ_A^+ is the first positive time the walk visits A. Good enough control of effective resistances allows one to understand most aspects of the random walk on a graph. We can also define effective resistances on infinite graphs, although issues arise with boundary conditions. An infinite graph is recurrent if and only if the effective resistance from a vertex to infinity is infinite.

The effective resistance can also be computed via either of two variational principles: *Dirichlet's principle* and *Thomson's principle*, see Sect. 2.4. The first expresses the effective resistance as a *supremum* of energies of a certain set of *functions*, while the second expresses the effective resistance as an *infimum* of energies of a certain set of *flows*. Thus, we can bound effective resistances from above by constructing flows, and from below by constructing functions. A central insight is that we can *use the circle packing* to construct these functions and flows. This idea leads fairly easily to various statements such as the following:

- The effective resistance across a Euclidean annulus of fixed modulus is at most a constant. If the triangulation has bounded degrees, then the resistance is at least a constant.

- The effective resistance between the left and right sides of a Euclidean square is at most a constant. If the triangulation has bounded degrees, then the resistance is at least a constant.

See for instance Lemma 4.9. We will use these ideas to prove the following remarkable theorem of He and Schramm [40], which pioneered the connection between circle packing and random walks.

Theorem 1.10 (He and Schramm 1995) *Let T be a one-ended infinite triangulation. If T has bounded degrees, then it is CP parabolic if and only if it is recurrent for simple random walk, that is, if and only if the simple random walk on T visits every vertex infinitely often almost surely.*

This has been extended to the multiply-ended cases in [32], see also Chap. 8, item 4.

Recurrence of Distributional Limits of Random Planar Maps

Random planar maps is a widely studied field lying at the intersection of probability, combinatorics and statistical physics. It aims to answer the vague question "what does a typical random surface look like?"

We provide here a very quick account of this field, referring the readers to the excellent lecture notes [58] by Le Gall and Miermont, and the many references within for further reading. The enumerative study of planar maps (answering questions of the form "how many simple triangulations on n vertices are there?") began with the work of Tutte in the 1960s [83] who enumerated various classes of finite planar maps, in particular triangulations. Cori and Vauquelin [18], Schaeffer [72] and Chassaing and Schaeffer [16] have found beautiful bijections between planar maps and labeled trees and initiated this fascinating topic in enumerative combinatorics. The bijections themselves are model dependent and extremely useful since many combinatorial and metric aspects of random planar maps can be inferred from them. This approach has spurred a new line of research: limits of large random planar maps.

Two natural notions of such limits come to mind: scaling limits and local limits. In the first notion, one takes a random planar map M_n on n vertices, scales the distances appropriately (in most models the correct scaling turns out to be $n^{-1/4}$), and aims to show that this random metric space converges in distribution in the Gromov-Hausdorff sense. The existence of such limits was suggested by Chassaing and Schaeffer [16], Le Gall [55], and Marckert and Mokkadem [62], who coined the term *the Brownian map* for such a limit. The recent landmark work of Le Gall [56] and Miermont [65] establishes the convergence of random p-angulations for $p = 3$ and all even p to the Brownian map.

The study of local limits of random planar maps, initiated by Benjamini and Schramm [11], while bearing many similarities, is independent of the study of scaling limits. The *local* limit of a random planar map M_n on n vertices is an infinite random rooted graph (U, ρ) with the property that neighborhoods of M_n around a random vertex converge in distribution to neighborhoods of U around ρ. The infinite random graph (U, ρ) captures the local behavior of M_n around typical vertices. We develop this notion precisely in Chap. 5.

In their pioneering work, Angel and Schramm [5] showed that the local limit of a uniformly chosen random triangulation on n vertices exists and that it is a one-ended infinite planar triangulation. They termed the limit as the *uniform infinite planar triangulation* (UIPT). The uniform infinite planar quadrangulation (UIPQ), that is, the local limit of a uniformly chosen random quadrangulation (i.e., each face has 4 edges) on n vertices, was later constructed by Krikun [52].

The questions in this line of research concern the almost sure properties of this limiting geometry. It is a highly fractal geometry that is drastically different from $\mathbb{Z}^2$. Angel [4] proved that the volume of a graph-distance ball of radius r in the UIPT is almost surely of order $r^{4+o(1)}$ and that the boundary component separating this ball from infinity has volume $r^{2+o(1)}$ almost surely. For the UIPQ this is proved in [16].

Due to the various combinatorial techniques of generating random planar maps, many of the metric properties of the UIPT/UIPQ are firmly understood. Surface properties of these maps are somewhat harder to understand using enumerative methods. Recall that a non-compact simply connected Riemannian surface is either conformally equivalent to the disc or the whole plane and that this is determined according to whether Brownian motion on the surface is transient or recurrent. Hence, the behavior of the simple random walk on the UIPT/UIPQ is considered here as a "surface property" (see also [30]).

As mentioned earlier, one of the main objectives of these notes is to answer the question of the almost sure recurrence/transience of the simple random walk on the UIPT/UIPQ. We provide a general statement, Theorem 6.1 of these notes, to which a corollary is

Theorem 1.11 ([31]) *The UIPT and UIPQ are almost surely recurrent.*

The proof heavily relies on the circle packing theorem and can be viewed as an extension of the remarkable theorem of Benjamini and Schramm [11] stating that the local limit of finite planar maps with finite maximum degree is almost surely recurrent. The maximum degree of the UIPT is unbounded and so one cannot apply [11]. A combination of the techniques presented in Chaps. 4–6 is required to overcome this difficulty.

Recently, there have been terrific new developments studying further surface properties of the UIPT/UIPQ. Lee [59] has given an exciting new proof of Theorem 1.11 based on a spectral analysis and an embedding theorem for planar maps due to [48]. His proof also yields that the spectral dimension of the UIPT/UIPQ is at most 2 and applies to local limits of sphere-packable graphs in higher dimensions as well. Gwynne and Miller [33] provided the converse bound showing that the spectral dimension of the UIPT equals 2 and calculated other exponents governing the behavior of the random walk. Their results are based on the deep work of Gwynne et al. [35] (see also Chap. 8, item 9).

Chapter 2
Random Walks and Electric Networks

An extremely useful tool and viewpoint for the study of random walks is Kirchhoff's theory of electric networks. Our treatment here roughly follows [69, Chapter 8], we also refer the reader to [61] for an in-depth comprehensive study.

Definition 2.1 A **network** is a connected graph $G = (V, E)$ endowed with positive edge weights, $\{c_e\}_{e \in E}$ (called **conductances**). The reciprocals $r_e = 1/c_e$ are called **resistances**.

In Sects. 2.1–2.4 below we discuss finite networks. We extend our treatment to infinite networks in Sect. 2.5.

2.1   Harmonic Functions and Voltages

Let $G = (V, E)$ be a finite network. In physics classes it is taught that when we impose specific voltages at fixed vertices a and z, then current flows through the network according to certain laws (such as the series and parallel laws). An immediate consequence of these laws is that the function from V to $\mathbb{R}$ giving the voltage at each vertex is harmonic at each $x \in V \setminus \{a, z\}$.

Definition 2.2 A function $h : V \to \mathbb{R}$ is **harmonic** at a vertex x if

$$h(x) = \frac{1}{\pi_x} \sum_{y:y \sim x} c_{xy} h(y) \qquad \text{where} \qquad \pi_x := \sum_{y:y \sim x} c_{xy}. \tag{2.1}$$

Instead of starting with the physical laws and proving that voltage is harmonic, we now take the axiomatically equivalent approach of defining voltage to be a harmonic function and deriving the laws as corollaries.

© The Author(s) 2020
A. Nachmias, *Planar Maps, Random Walks and Circle Packing*, Lecture Notes in Mathematics 2243, https://doi.org/10.1007/978-3-030-27968-4_2

Definition 2.3 Given a network $G = (V, E)$ and two distinct vertices $a, z \in V$, a **voltage** is a function $h : V \to \mathbb{R}$ that is harmonic at any $x \in V \setminus \{a, z\}$.

We will show in Claim 2.8 and Corollary 2.7 that for any $\alpha, \beta \in \mathbb{R}$, there is a unique voltage h such that $h(a) = \alpha$ and $h(z) = \beta$ (this assertion is true only when the network is finite).

Claim 2.4 If h_1, h_2 are harmonic at x then so is any linear combination of h_1, h_2.

Proof Let $\bar{h} = \alpha h_1 + \beta h_2$ for some $\alpha, \beta \in \mathbb{R}$. It holds that

$$\bar{h}(x) = \alpha h_1(x) + \beta h_2(x) = \frac{1}{\pi_x} \sum_{y:y \sim x} c_{xy} \alpha h_1(y) + \frac{1}{\pi_x} \sum_{y:y \sim x} c_{xy} \beta h_2(y)$$

$$= \frac{1}{\pi_x} \sum_{y:y \sim x} c_{xy} \bar{h}(y). \qquad \qquad \square$$

Claim 2.5 If $h : V \to \mathbb{R}$ is harmonic at all the vertices of a finite network, then it is constant.

Proof Let $M = \sup_x h(x)$ be the maximum value of h. Let $A = \{x \in V : h(x) = M\}$. Since G is finite, $A \neq \emptyset$. Given $x \in A$, we have that $h(y) \leq h(x)$ for all neighbors y of x. By harmonicity, $h(x)$ is the weighted average of the values of $h(y)$ at the neighbors; but this can only happen if all neighbors of x are also in A. Since G is connected we obtain that $A = V$ implying that h is constant. $\qquad \square$

We now show that a voltage is determined by its boundary values, i.e., by its values at a, z.

Claim 2.6 If h is a voltage satisfying $h(a) = h(z) = 0$, then $h \equiv 0$.

Proof Put $M = \max_x h(x)$ (which is attained since G is finite) and let $A = \{x \in V : h(x) = M\}$. As before, by harmonicity, if $x \in A \setminus \{a, z\}$ then all of its neighbors are also in A. Since G is connected, there exists a simple path from x to either a or z such that only its endpoint is in $\{a, z\}$. Since $h(a) = h(z) = 0$ we learn that $M = 0$, that is, h is non-positive. Similarly, one proves that h is non-negative, thus $h \equiv 0$. $\square$

Corollary 2.7 (Voltage Uniqueness) *For every $\alpha, \beta \in \mathbb{R}$, if h, h' are voltages satisfying $h(a) = h'(a) = \alpha$ and $h(z) = h'(z) = \beta$, then $h \equiv h'$.*

Proof By Claim 2.4, the function $h - h'$ is a voltage, taking the value 0 at a and z, hence by Claim 2.6 we get $h \equiv h'$. $\qquad \square$

Claim 2.8 For every $\alpha, \beta \in \mathbb{R}$, there exists a voltage h satisfying $h(a) = \alpha, h(z) = \beta$.

Proof 1 We write $n = |V|$. Observe that a voltage h with $h(a) = \alpha$ and $h(z) = \beta$ is defined by a system of $n - 2$ linear equations of the form (2.1) in $n - 2$ variables (which are the values $h(x)$ for $x \in V \setminus \{a, z\}$). Corollary 2.7 guarantees that the matrix representing that system has empty kernel, hence it is invertible. $\qquad \square$

We present an alternative proof of existence based on the random walk on the network. Consider the Markov chain $\{X_n\}$ on the state space V with transition probabilities

$$p_{xy} := \mathbb{P}(X_{t+1} = y \mid X_t = x) = \frac{c_{xy}}{\pi_x}. \tag{2.2}$$

This Markov chain is a **weighted random walk** (note that if c_{xy} are all 1 then the described chain is the so-called **simple random walk**). We write $\mathbb{P}_x$ and $\mathbb{E}_x$ for the probability and expectation, respectively, conditioned on $X_0 = x$. For a vertex x, define the **hitting time** of x by

$$\tau_x := \min\{t \geq 0 \mid X_t = x\}.$$

Proof 2 We will find a voltage g satisfying $g(a) = 0$ and $g(z) = 1$ by setting

$$g(x) = \mathbb{P}_x(\tau_z < \tau_a).$$

Indeed, g is harmonic at $x \neq a, z$, since by the law of total probability and the Markov property we have

$$g(x) = \frac{1}{\pi_x} \sum_{y:y\sim x} c_{xy} \mathbb{P}_x(\tau_z < \tau_a \mid X_1 = y) = \frac{1}{\pi_x} \sum_{y:y\sim x} c_{xy} \mathbb{P}_y(\tau_z < \tau_a)$$

$$= \frac{1}{\pi_x} \sum_{y:y\sim x} c_{xy} g(y).$$

For general boundary conditions α, β we define h by

$$h(x) = g(x) \cdot (\beta - \alpha) + \alpha.$$

By Claim 2.4, h is a voltage, and clearly $h(a) = \alpha$ and $h(z) = \beta$, concluding the proof. $\qquad \square$

This proof justifies the equality between simple random walk probabilities and voltages that was discussed at the start of this chapter: since the function $x \mapsto \mathbb{P}_x(\tau_z < \tau_a)$ is harmonic on $V \setminus \{a, z\}$ and takes values 0, 1 at a, z respectively, it must be equal to the voltage at x when voltages 0, 1 are imposed at a, z.

Claim 2.9 If h is a voltage with $h(a) \leq h(z)$, then $h(a) \leq h(x) \leq h(z)$ for all $x \in V$.

Furthermore, if $h(a) < h(z)$ and $x \in V \setminus \{a, z\}$ is a vertex such that x is in the connected component of z in the graph $G \setminus \{a\}$, and x is in the connected component of a in the graph $G \setminus \{z\}$, then $h(a) < h(x) < h(z)$.

Proof This follows directly from the construction of h in Proof 2 of Claim 2.8 and the uniqueness statement of Corollary 2.7. Alternatively, one can argue as in the proof of Claim 2.6 that if $M = \max_x h(x)$ and $m = \min_x h(x)$, then the sets $A = \{x \in V : h(x) = M\}$ and $B = \{x \in V : h(x) = m\}$ must each contain at least one element of $\{a, z\}$.

To prove the second assertion, we note that by Claim 2.8 and Corollary 2.7 it is enough to check when h is the voltage with boundary values $h(a) = 0$ and $h(z) = 1$. In this case, the condition on x guarantees that the probabilities that the random walk started at x visits a before z or visits z before a are positive. By proof 2 of Claim 2.8 we find that $h(x) \in (0, 1)$.   □

2.2   Flows and Currents

For a graph $G = (V, E)$, denote by $\vec{E}$ the set of edges of G, each endowed with the two possible orientations. That is, $(x, y) \in \vec{E}$ iff $\{x, y\} \in E$ (and in that case, $(y, x) \in \vec{E}$ as well).

Definition 2.10 A **flow from** a **to** z in a network G is a function $\theta : \vec{E} \to \mathbb{R}$ satisfying

1. For any $\{x, y\} \in E$ we have $\theta(xy) = -\theta(yx)$ (**antisymmetry**), and
2. $\forall x \notin \{a, z\}$ we have $\sum_{y:y \sim x} \theta(xy) = 0$ (**Kirchhoff's node law**).

Claim 2.11 If θ_1, θ_2 are flows then, so is any linear combination of θ_1, θ_2.

Proof Let $\bar{\theta} = \alpha\theta_1 + \beta\theta_2$ for some $\alpha, \beta \in \mathbb{R}$. It holds that

$$\bar{\theta}(xy) = \alpha\theta_1(xy) + \beta\theta_2(xy) = -\alpha\theta_1(yx) - \beta\theta_2(yx) = -\bar{\theta}(yx),$$

and for $x \neq a, z$,

$$\sum_{y:y \sim x} \bar{\theta}(xy) = \alpha \sum_{y:y \sim x} \theta_1(xy) + \beta \sum_{y:y \sim x} \theta_2(xy) = 0. \qquad □$$

Definition 2.12 Given a voltage h, the **current flow** $\theta = \theta_h$ associated with h is defined by $\theta(xy) = c_{xy}(h(y) - h(x))$.

In other words, the voltage difference across an edge is the product of the current flowing along the edge with the resistance of the edge. This is known as **Ohm's law**. According to this definition, the current flows from vertices with lower voltage to vertices with higher voltage. We will use this convention throughout, but the reader should be advised that some other sources use the opposite convention.

Claim 2.13 The current flow associated with a voltage is indeed a flow.

Proof The current flow is clearly antisymmetric by definition. To show that it satisfies the node law, observe that for $x \neq a, z$, since h is harmonic,

$$\sum_{y:y\sim x} \theta(xy) = \overbrace{\sum_{y:y\sim x} c_{xy}h(y)}^{=\pi_x h(x)} - \overbrace{\sum_{y:y\sim x} c_{xy}h(x)}^{=\pi_x h(x)} = 0. \qquad \square$$

Claim 2.14 The current flow associated with a voltage h satisfies **Kirchhoff's cycle law**, that is, for every directed cycle $\vec{e}_1, \ldots, \vec{e}_m$,

$$\sum_{i=1}^{r} r_{e_i}\theta(\vec{e}_i) = 0.$$

Proof Write $\vec{e}_i = (x_{i-1}, x_i)$, and observe that $x_0 = x_m$. We have that

$$\sum_{i=1}^{m} r_{e_i}\theta(\vec{e}_i) = \sum_{i=1}^{m} r_{x_{i-1}x_i} c_{x_{i-1}x_i}(h(x_i) - h(x_{i-1})) = \sum_{i=1}^{m}(h(x_i) - h(x_{i-1})) = 0. \qquad \square$$

For examples of a flow which does not satisfy the cycle law and a current flow, see Fig. 2.1.

Claim 2.15 Given a flow θ which satisfies the cycle law, there exists a voltage $h = h_\theta$ such that θ is the current flow associated with h. Furthermore, this voltage is unique up to an additive constant.

Proof For every vertex x, let $\vec{e}_1, \ldots, \vec{e}_k$ be a path from a to x, and define

$$h(x) = \sum_{i=1}^{k} r_{e_i}\theta(\vec{e}_i). \qquad (2.3)$$

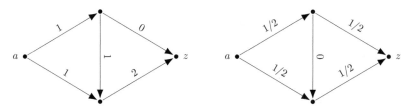

Fig. 2.1 On the left, a flow of strength 2 in which the cycle law is violated. On the right, the unit (i.e., strength 1) current flow

Note that since θ satisfies the cycle law, the right hand side of (2.3) does not depend on the choice of the path, hence $h(x)$ is well defined. Let $x \in V$, and consider a given path $\vec{e}_1, \ldots, \vec{e}_k$ from a to x (if $x = a$ we take the empty path). To evaluate $h(y)$ for $y \sim x$, consider the path $\vec{e}_1, \ldots, \vec{e}_k, xy$ from a to y, so $h(y) = h(x) + r_{xy}\theta(xy)$. It follows that $h(y) - h(x) = r_{xy}\theta(xy)$, hence $\theta(xy) = c_{xy}(h(y) - h(x))$, meaning that θ is indeed the current flow associated with h.

Since $\theta(xy) = c_{xy}(h(y) - h(x))$ for any $x \sim y$, the node law of immediately implies that h is a voltage. To show that h is unique up to an additive constant, suppose that $g : V \to \mathbb{R}$ is another voltage such that $r_{xy}\theta(xy) = g(y) - g(x)$. It follows that $g(y) - h(y) = g(x) - h(x)$ for any $x \sim y$. Since G is connected it follows that $g - h$ is the constant function on V. $\square$

Definition 2.16 The **strength** of a flow θ is

$$\|\theta\| = \sum_{x:x \sim a} \theta(ax).$$

Claim 2.17 For every flow θ,

$$\sum_{x:x \sim z} \theta(xz) = \|\theta\|.$$

Proof We have that

$$0 = \sum_{x \in V} \sum_{y:y \sim x} \theta(xy)$$

$$= \sum_{x \in V \setminus \{a,z\}} \sum_{y:y \sim x} \theta(xy) + \sum_{y:y \sim a} \theta(ay) + \sum_{y:y \sim z} \theta(zy)$$

$$= \sum_{y:y \sim a} \theta(ay) + \sum_{y:y \sim z} \theta(zy)$$

where the first equality is due to antisymmetry, and the third equality is due to the node law. The claim follows again by antisymmetry. $\square$

Claim 2.18 If θ_1, θ_2 are flows satisfying the cycle law and $\|\theta_1\| = \|\theta_2\|$, then $\theta_1 = \theta_2$.

Proof Let $\bar{\theta} = \theta_1 - \theta_2$. According to Claim 2.11, $\bar{\theta}$ is a flow. It also satisfies the cycle law, as for every cycle $\vec{e}_1, \ldots, \vec{e}_m$,

$$\sum_{i=1}^{m} r_{e_i} \bar{\theta}(\vec{e}_i) = \sum_{i=1}^{m} r_{e_i} \theta_1(\vec{e}_i) - \sum_{i=1}^{m} r_{e_i} \theta_2(\vec{e}_i) = 0.$$

Observe in addition that $\|\bar{\theta}\| = \|\theta_1\| - \|\theta_2\| = 0$. Now, let $h = h_{\bar{\theta}}$ be the voltage defined in Claim 2.15, chosen so that $h(a) = 0$. Note that it is harmonic at a, since

$$\frac{1}{\pi_a} \sum_{x:x\sim a} c_{ax} h(x) = \frac{1}{\pi_a} \sum_{x:x\sim a} c_{ax}(h(a) + r_{ax}\bar{\theta}(ax))$$

$$= \frac{1}{\pi_a} \sum_{x:x\sim a} c_{ax} h(a) + \frac{1}{\pi_a} \sum_{x:x\sim a} \bar{\theta}(ax) = h(a) + \frac{\|\bar{\theta}\|}{\pi_a} = h(a).$$

Similarly, using Claim 2.17 it is also harmonic at z. Since h is harmonic everywhere, it is constant by Claim 2.5, and thus $h \equiv 0$, hence $\bar{\theta} \equiv 0$ and so $\theta_1 = \theta_2$.   □

This last claim prompts the following useful definition.

Definition 2.19 The **unit current flow** from a to z is the unique current flow from a to z of strength 1.

2.3   The Effective Resistance of a Network

Suppose we are given a voltage h on a network G with fixed vertices a and z. Scaling h by a constant multiple causes the associated current flow to scale by the same multiple, while adding a constant to h does not change the current flow at all. Therefore, the strength of the current flow is proportional to the difference $h(z) - h(a)$.

Claim 2.20 For every non-constant voltage h and a current flow θ corresponding to h, the ratio

$$\frac{h(z) - h(a)}{\|\theta\|} \tag{2.4}$$

is a positive constant which does not depend on h.

Proof Let h_1, h_2 be two non-constant voltages, and let θ_1, θ_2' be their associated current flows. For $i = 1, 2$, let $\bar{h}_i = h_i / \|\theta_i\|$ and let $\bar{\theta}_i$ be the current flow associated with $\bar{h}_i$ (note that since h_i is non-constant $\|\theta_i\| \neq 0$). Thus, $\|\bar{\theta}_i\| = 1$. By Claim 2.18 we get $\bar{\theta}_1 = \bar{\theta}_2$ and therefore $\bar{h}_1 = \bar{h}_2 + c$ for some constant c by Claim 2.15. It follows that $\bar{h}_1(z) - \bar{h}_1(a) = \bar{h}_2(z) - \bar{h}_2(a)$.

To see that this constant is positive, it is enough to check one particular choice of a voltage. By Claim 2.8, let h be the voltage with $h(a) = 0$ and $h(z) = 1$. By Claim 2.9 and since G is connected, we have that $h(x) > 0$ for at least one neighbor x of a. Thus, the corresponding current flow θ has $\|\theta\| > 0$ making (2.4) positive.
□

Fig. 2.2 Examples for effective resistances of two networks with unit edge conductances. (a) For the voltage depicted, the voltage difference between a and z is 5, and the current flow's strength is 1, hence the effective resistance is $5/1 = 5$. (b) For the voltage depicted, the voltage difference between a and z is 1, and the current flow's strength is 1, hence the effective resistance is $1/1 = 1$

Claim 2.20 is the mathematical manifestation of Ohm's law which states that the voltage difference across an electric circuit is proportional to the current through it. The constant of proportionality is usually called the *effective resistance* of the circuit.

Definition 2.21 The number defined in (2.4) is called the **effective resistance** between a and z in the network, and is denoted $\mathcal{R}_{\mathrm{eff}}(a \leftrightarrow z)$. We call its reciprocal the **effective conductance** between a and z and is denoted $\mathcal{C}_{\mathrm{eff}}(a \leftrightarrow z) := \mathcal{R}_{\mathrm{eff}}(a \leftrightarrow z)^{-1}$.

For examples of computing the effective resistances of networks, see Fig. 2.2.

Notation In most cases we write $\mathcal{R}_{\mathrm{eff}}(a \leftrightarrow z)$ and suppress the notation of which network we are working on. However, when it is important to us what the network is, we will write $\mathcal{R}_{\mathrm{eff}}(a \leftrightarrow z; G)$ for the effective resistance in the network G with unit edge conductances and $\mathcal{R}_{\mathrm{eff}}(a \leftrightarrow z; (G, \{r_e\}))$ for the effective resistance in the network G with edge resistances $\{r_e\}_{e \in E}$. Furthermore, given disjoint subsets A and Z of vertices in a graph G, we write $\mathcal{R}_{\mathrm{eff}}(A \leftrightarrow Z)$ for the effective resistance between a and z in the network obtained from the original network by identifying all the vertices of A into a single vertex a, and all the vertices of Z into a single vertex z.

Probabilistic Interpretation For a vertex x we write τ_x^+ for the stopping time

$$\tau_x^+ = \min\{t \geq 1 \mid X_t = x\}, \tag{2.5}$$

where X_t is the weighted random walk on the network, as defined in (2.2). Note that if $X_0 \neq x$ then $\tau_x = \tau_x^+$ with probability 1.

Claim 2.22

$$\mathcal{R}_{\mathrm{eff}}(a \leftrightarrow z) = \frac{1}{\pi_a \mathbb{P}_a(\tau_z < \tau_a^+)}.$$

Proof Consider the voltage h satisfying $h(a) = 0$ and $h(z) = 1$, and let θ be the current flow associated with h. Due to uniqueness of h (Corollary 2.7) we have that

for $x \neq a, z$,

$$h(x) = \mathbb{P}_x(\tau_z < \tau_a),$$

hence

$$\mathbb{P}_a(\tau_z < \tau_a^+) = \frac{1}{\pi_a} \sum_{x \sim a} c_{ax} \mathbb{P}_x(\tau_z < \tau_a)$$

$$= \frac{1}{\pi_a} \sum_{x \sim a} c_{ax} h(x)$$

$$= \frac{1}{\pi_a} \sum_{x \sim a} \theta(ax) = \frac{\|\theta\|}{\pi_a} = \frac{1}{\pi_a \mathcal{R}_{\mathrm{eff}}(a \leftrightarrow z)}. \qquad \square$$

Network Simplifications   Sometimes a network can be replaced by a simpler network, without changing the effective resistance between a pair of vertices.

Claim 2.23 (Parallel Law)   Conductances add in parallel. Suppose e_1, e_2 are parallel edges between a pair of vertices, with conductances c_1 and c_2, respectively. If we replace them with a single edge e' with conductance $c_1 + c_2$, then the effective resistance between a and z is unchanged.

A demonstration of the parallel law appears in Fig. 2.3.

Proof Let G' be the graph where e_1 and e_2 are replaced with e' with conductance $c_1 + c_2$. Then it is immediate that if h is any voltage function on G, then it remains a voltage function on the network G'. The claim follows. $\square$

Claim 2.24 (Series Law)   Resistances add in series. Suppose that $u \notin \{a, z\}$ is a vertex of degree 2 and that $e_1 = (u, v_1)$ and $e_2 = (u, v_2)$ are the two edges touching u with edge resistances r_1 and r_2, respectively. If we erase u and replace e_1 and e_2 by a single edge $e' = (v_1, v_2)$ of resistance $r_1 + r_2$, then the effective resistance between a and z is unchanged.

The series law is depicted in Fig. 2.4.

Proof Denote by G' the graph in which u is erased and e_1 and e_2 are replaced by a single edge (v_1, v_2) of resistance $r_1 + r_2$. Let θ be a current flow from a to z in G, and define a flow θ' from a to z in G' by putting $\theta'(e) = \theta(e)$ for any $e \neq e_1, e_2$ and $\theta'(v_1, v_2) = \theta(v_1, u)$. Since u had degree 2, it must be that $\theta(v_1, u) = \theta(u, v_2)$.

Fig. 2.3 Demonstrating the parallel law. Two parallel edges are replaced by a single edge

$$v_1 \bullet \xrightarrow{\quad r_1 \quad} \overset{u}{\bullet} \xrightarrow{\quad r_2 \quad} \bullet\, v_2 \qquad\qquad v_1 \bullet \xrightarrow{\quad r_1 + r_2 \quad} \bullet\, v_2$$

Fig. 2.4   An example of a network G where edges in series are replaced by a single edge

Thus θ' satisfies the node law at any $x \notin \{a, z\}$ and $\|\theta\| = \|\theta'\|$. Furthermore, since θ satisfies the cycle law, so does θ'. We conclude θ' is a current flow of the same strength as θ and the voltage difference they induce is the same.   □

The operation of **gluing** a subset of vertices $S \subset V$ consists of identifying the vertices of S into a single vertex and keeping all the edges and their conductances. In this process we may generate parallel edges or loops.

Claim 2.25 (Gluing)   Gluing vertices of the same voltage does not change the effective resistance between a and z.

Proof   This is immediate since the voltage on the glued graph is still harmonic.   □

Example: Spherically Symmetric Tree Let Γ be a **spherically symmetric tree**, that is, a rooted tree where all vertices at the same distance from the root have the same number of children. Denote by ρ the root of the tree, and let $\{d_n\}_{n\in\mathbb{N}}$ be a sequence of positive integers. Every vertex at distance n from the root ρ has d_n children. Denote by Γ_n the set of all vertices of height n. We would like to calculate $\mathcal{R}_{\text{eff}}(\rho \leftrightarrow \Gamma_n)$. Due to the tree's symmetry, all vertices at the same level have the same voltage and therefore by Claim 2.25 we can identify them. Our simplified network has now one vertex for each level, denoted by $\{v_i\}_{i\in\mathbb{N}}$ (where $\rho = v_0$), with $|\Gamma_{n+1}|$ edges between v_n and v_{n+1}. Using the parallel law (Claim 2.23), we can reduce each set of $|\Gamma_n|$ edges to a single edge with resistance $\frac{1}{|\Gamma_n|}$, then, using the series law (Claim 2.24) we get

$$\mathcal{R}_{\text{eff}}(\rho \leftrightarrow \Gamma_n) = \sum_{i=1}^{n} \frac{1}{|\Gamma_i|} = \sum_{i=1}^{n} \frac{1}{d_0 \cdots d_{i-1}} \,,$$

see Fig. 2.5.

By Claim 2.22 we learn that

$$\mathbb{P}_\rho(\tau_n < \tau_\rho^+) = \frac{1}{d_0 \sum_{i=1}^{n} \frac{1}{d_0 \cdots d_{i-1}}}, \tag{2.6}$$

where τ_n is the hitting time of Γ_n for the random walk on Γ. Observe that

$$\mathbb{P}_\rho\left(\tau_n < \tau_\rho^+ \text{ for all } n\right) = \mathbb{P}_\rho\left(X_t \text{ never returns to } \rho\right) \,,$$

Fig. 2.5 Using network simplifications. (**a**) The first four levels of a spherically symmetric tree with $\{d_n\} = \{3, 2, 1, \ldots\}$. (**b**) Gluing nodes on the same level. (**c**) Applying the parallel law. (**d**) Applying the series law

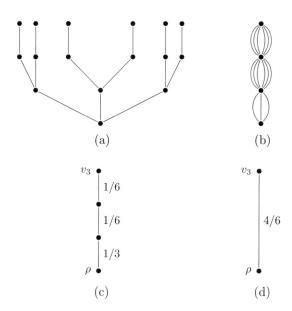

so by (2.6) we reach an interesting dichotomy. If $\sum_{i=1}^{\infty} \frac{1}{d_1 \cdots d_i} = \infty$, then the random walker returns to ρ with probability 1, and hence returns to ρ infinitely often almost surely. If $\sum_{i=1}^{\infty} \frac{1}{d_1 \cdots d_i} < \infty$, then with positive probability the walker never returns to ρ, and hence visits ρ only finitely many times almost surely.

The former graph is called a **recurrent** graph and the latter is called **transient**. We will get back to this dichotomy in Sect. 2.5.

The Commute Time Identity

The following lemma shows that the effective resistance between a and z is proportional to the expected time it takes the random walk starting at a to visit z and then return to a, in other words, the expected *commute time* between a and z. We will use this lemma only in Chap. 6 so the impatient reader may skip this section and return to it later.

Lemma 2.26 (Commute Time Identity)   *Let $G = (V, E)$ be a finite network and $a \neq z$ two vertices. Then*

$$\mathbb{E}_a[\tau_z] + \mathbb{E}_z[\tau_a] = 2\mathcal{R}_{\text{eff}}(a \leftrightarrow z) \sum_{e \in E} c_e$$

Proof We denote by $G_z : V \times V \to \mathbb{R}$ the so-called Green function

$$G_z(a, x) = \mathbb{E}_a[\text{number of visits to } x \text{ before } z]$$

and note that

$$\mathbb{E}_a[\tau_z] = \sum_{x \in V} G_z(a, x).$$

It is straightforward to show that the function $v(x) = G_z(a, x)/\pi_x$ is harmonic in $V \setminus \{a, z\}$. Also, we have that $G_z(a, z) = 0$ and $G_z(a, a) = \frac{1}{\mathbb{P}_a(\tau_z < \tau_a)} = \pi_a \mathcal{R}_{\text{eff}}(a \leftrightarrow z)$. Thus, v is a voltage function with boundary conditions $v(z) = 0$ and $v(a) = \mathcal{R}_{\text{eff}}(a \leftrightarrow z)$ which satisfies

$$\mathbb{E}_a[\tau_z] = \sum_{x \in V} v(x)\pi_x \,.$$

Similarly, the same analysis for $\mathbb{E}_z[\tau_a]$ yields the same result, with the voltage function η which has boundary conditions $\eta(z) = \mathcal{R}_{\text{eff}}(a \leftrightarrow z)$ and $\eta(a) = 0$. Therefore, $\eta(x) = v(a) - v(x)$ for all $x \in V$ since both sides are harmonic functions in $V \setminus \{a, z\}$ that receive the same boundary values. This implies that

$$\mathbb{E}_z[\tau_a] = \sum_{x \in V} \pi_x \left(v(a) - v(x) \right).$$

Summing these up gives

$$\mathbb{E}_a[\tau_z] + \mathbb{E}_z[\tau_a] = \sum_{x \in V} \pi_x v(a) = 2 \sum_{e \in E} c_e \mathcal{R}_{\text{eff}}(a \leftrightarrow z). \qquad \square$$

2.4 Energy

So far we have seen how to compute the effective resistance of a network via harmonic functions and current flows. However, in typical situations it is hard to find a flow satisfying the circle law. Luckily, an extremely useful property of the effective resistance is that it can be represented by a variational problem. Our intuition from highschool physics suggests that the *energy* of the unit current flow is minimal among all unit flows from a to z. The notion of energy can be made precise and will allow us to obtain valuable monotonicity properties. For instance, removing any edge from an electric network can only increase its effective resistance. Hence, any recurrent graph remains recurrent after removing any subset of edges from it. Two variational problems govern the effective resistance, Thomson's principle, which is used to bound the effective resistance from above, and Dirichlet's principle, used to bound it from below.

Definition 2.27 The **energy** of a flow θ from a to z, denoted by $\mathcal{E}(\theta)$, is defined to be

$$\mathcal{E}(\theta) := \frac{1}{2} \sum_{\vec{e} \in \vec{E}} r_{\vec{e}} \theta(\vec{e})^2 = \sum_{e \in E} \theta(e)^2 r_e.$$

Note that in the second sum we sum over undirected edges, but since $\theta(xy)^2 = \theta(yx)^2$, this is well defined.

Theorem 2.28 (Thomson's Principle)

$$\mathcal{R}_{\text{eff}}(a \leftrightarrow z) = \inf\{\mathcal{E}(\theta) : \|\theta\| = 1, \theta \text{ is a flow from } a \text{ to } z\}$$

and the unique minimizer is the unit current flow.

Proof First, we will show that the energy of the unit current flow is the effective resistance. Let I be the unit current flow, and h the corresponding (Claim 2.15) voltage function.

$$\mathcal{E}(I) = \frac{1}{2} \sum_{x \in V} \sum_{y:y \sim x} r_{xy} I(xy)^2 = \frac{1}{2} \sum_{x \in V} \sum_{y:y \sim x} r_{xy} \left(\frac{h(y) - h(x)}{r_{xy}} \right) I(xy)$$

$$= \frac{1}{2} \sum_{x \in V} \sum_{y:y \sim x} (h(y) - h(x)) I(xy)$$

$$= \frac{1}{2} \sum_{x \in V} \sum_{y:y \sim x} h(y) I(xy) - \frac{1}{2} \sum_{x \in V} \sum_{y:y \sim x} h(x) I(xy).$$

Observe that in the second term of the right hand side, for every $x \neq a, z$ the sum over all $y \sim x$ is 0 due to the node law, hence the entire term equals $\frac{1}{2}(h(a) - h(z))$. From antisymmetry of I, the first term on the right hand side equals $-\frac{1}{2}(h(a) - h(z))$, hence the right hand side equals altogether $h(z) - h(a) = \mathcal{R}_{\text{eff}}(a \leftrightarrow z)$.

We will now show that every other flow J with $\|J\| = 1$ has $\mathcal{E}(J) \geq \mathcal{E}(I)$. Let J be such flow and write $J = I + (J - I)$. Set $\theta = J - I$ and note that $\|\theta\| = 0$. We have

$$\mathcal{E}(J) = \frac{1}{2} \sum_{x \in V} \sum_{y:y \sim x} r_{xy} (I(xy) + \theta(xy))^2$$

$$= \frac{1}{2} \sum_{x \in V} \sum_{y:y \sim x} r_{xy} I(xy)^2 + \frac{1}{2} \sum_{x \in V} \sum_{y:y \sim x} r_{xy} \theta(xy)^2$$

$$+ \sum_{x \in V} \sum_{y:y \sim x} r_{xy} \theta(xy) I(xy)$$

$$= \mathcal{E}(I) + \mathcal{E}(\theta) + \sum_{x \in V} \sum_{y:y \sim x} r_{xy} \theta(xy) I(xy).$$

Now,

$$\sum_{x \in V} \sum_{y:y \sim x} r_{xy} \theta(xy) I(xy) = \sum_{x \in V} \sum_{y:y \sim x} r_{xy} \theta(xy) \frac{(h(y) - h(x))}{r_{xy}}$$

$$= \sum_{x \in V} \sum_{y:y \sim x} \theta(xy) (h(y) - h(x))$$

$$= 2 \cdot \|\theta\| \cdot (h(z) - h(a)) = 0,$$

where the last inequality follows from the same reasoning as before. We conclude that $\mathcal{E}(J) \geq \mathcal{E}(I)$ as required and that equality holds if and only if $\mathcal{E}(\theta) = 0$, that is, if and only if $J = I$.   □

Corollary 2.29 (Rayleigh's Monotonicity Law) *If $\{r_e\}_{e \in E}$ and $\{r'_e\}_{e \in E}$ are edge resistances on the same graph G so that $r_e \leq r'_e$ for all edges $e \in E$, then*

$$\mathcal{R}_{\text{eff}}(a \leftrightarrow z; (G, \{r_e\})) \leq \mathcal{R}_{\text{eff}}(a \leftrightarrow z; (G, \{r'_e\})).$$

Proof Let θ be a flow on G, then

$$\sum_{e \in E} r_e \theta(e)^2 \leq \sum_{e \in E} r'_e \theta(e)^2.$$

This inequality is preserved while taking infimum over all flows with strength 1. Applying Theorem 2.28 finishes the proof.   □

Corollary 2.30 *Gluing vertices cannot increase the effective resistance between a and z.*

Proof Denote by G the original network and by G' the network obtained from gluing a subset of vertices. Then every flow θ on G (viewed as a function on the edges) is a flow on G'. Hence the infimum in Theorem 2.28 taken over flows in G' is taken over a larger subset of flows.   □

Definition 2.31 The **energy** of a *function* $h : V \to \mathbb{R}$, denoted by $\mathcal{E}(h)$, is defined to be

$$\mathcal{E}(h) := \sum_{\{x,y\} \in E} c_{xy} (h(x) - h(y))^2.$$

Compare the following lemma with Thomson's principle (Theorem 2.28).

Lemma 2.32 (Dirichlet's Principle) *Let G be a finite network with source a and sink z. Then*

$$\frac{1}{\mathcal{R}_{\text{eff}}(a \leftrightarrow z)} = \inf \big\{ \mathcal{E}(h) : h : V \to \mathbb{R}, \, h(a) = 0, \, h(z) = 1 \big\}.$$

Proof The infimum is obtained when h is *the* harmonic function taking 0 and 1 at a, z respectively. The reason is that if there exists $v \neq a, z$ with

$$h(v) \neq \sum_{u \sim v} \frac{c_{vu}}{\pi_v} h(u), \tag{2.7}$$

then we can change the value of h at v to be the right hand side of (2.7) and the energy will only decrease. One way to see this is that if X is a random variable with a second moment, then the value $\mathbb{E}(X)$ minimizes the function $f(x) = \mathbb{E}\left((X - x)^2\right)$.

Let h be that harmonic function and let I be its current flow, so $I(xy) = c_{xy}(h(y) - h(x))$. Write $\hat{I} = \mathcal{R}_{\text{eff}}(a \leftrightarrow z) \cdot I$, so $\|\hat{I}\| = 1$. By Thomson's principle,

$$\mathcal{R}_{\text{eff}}(a \leftrightarrow z) = \mathcal{E}(\hat{I}) = \sum_{e \in E} r_e \hat{I}(e)^2 = \sum_{\{x,y\} \in E} r_{xy} \mathcal{R}_{\text{eff}}(a \leftrightarrow z)^2 c_{xy}^2 (h(y) - h(x))^2,$$

hence

$$\frac{1}{\mathcal{R}_{\text{eff}}(a \leftrightarrow z)} = \mathcal{E}(h). \qquad \square$$

2.5 Infinite Graphs

Let $G = (V, E)$ be an infinite connected graph with edge resistances $\{r_e\}_{e \in E}$. We assume henceforth that this network is **locally finite**, that is, for any vertex $x \in V$ we have $\sum_{y:y \sim x} c_{xy} < \infty$. Let $\{G_n\}$ be a sequence of finite subgraphs of G such that $\bigcup_{n \in \mathbb{N}} G_n = G$ and $G_n \subset G_{n+1}$; we call such a sequence an **exhaustive sequence** of G. Identify all vertices of $G \setminus G_n$ with a single vertex z_n.

Claim 2.33 Given an exhaustive sequence $\{G_n\}$ of G, the limit

$$\lim_{n \to \infty} \mathcal{R}_{\text{eff}}(a \leftrightarrow z_n; G_n \cup \{z_n\}) \tag{2.8}$$

exists.

Proof The graph $G_n \cup \{z_n\}$ can be obtained from $G_{n+1} \cup \{z_{n+1}\}$ by gluing the vertices in $G_{n+1} \setminus G_n$ with z_{n+1} and labeling the new vertex z_n. By Corollary 2.30, the effective resistance $\mathcal{R}_{\text{eff}}(a \leftrightarrow z_n; G_n \cup \{z_n\})$ is increasing in n. $\qquad \square$

Claim 2.34 The limit in (2.8) does not depend on the choice of exhaustive sequence $\{G_n\}$.

Proof Indeed, let $\{G_n\}$ and $\{G'_n\}$ be two exhaustive sequences of G. We can find subsequences $\{i_k\}_{k\geq 1}$ and $\{j_k\}_{k\geq 1}$ such that

$$G_{i_1} \subseteq G'_{j_1} \subseteq G_{i_2} \subseteq \dots$$

Since $\{G_{i_1}, G'_{j_1}, G_{i_2}, \dots\}$ is itself an exhaustive sequence of G, the limit of effective resistances for this sequence exists and equals the limits of effective resistances for the subsequences $\{G_{i_k}\}$ and $\{G'_{j_k}\}$. In turn, these are equal to the limits of effective resistances for the original sequences $\{G_n\}$ and $\{G'_n\}$, respectively.   □

Definition 2.35 In an infinite network, the effective resistance from a vertex a and ∞ is

$$\mathcal{R}_{\text{eff}}(a \leftrightarrow \infty) := \lim_{n\to\infty} \mathcal{R}_{\text{eff}}(a \leftrightarrow z_n; G_n \cup \{z_n\}) .$$

We are now able to address the question of recurrence versus transience of a graph systematically. Recall the definition of τ_x^+ in (2.5). In an infinite network we define $\tau_a^+ = \infty$ when there is no time $t \geq 1$ such that $X_t = a$.

Definition 2.36 A network $(G, \{r_e\}_{e\in E})$ is called **recurrent** if $\mathbb{P}_a(\tau_a^+ = \infty) = 0$, that is, if the probability of the random walker started at a never returning to a is 0. Otherwise, it is called **transient** .

Observe that since G is connected, if $\mathbb{P}_a(\tau_a^+ = \infty) = 0$ for one vertex a, then it holds for all vertices in the network. As we have seen, if n is large enough so that $a \in G_n$, then

$$\mathcal{R}_{\text{eff}}(a \leftrightarrow z_n; G_n \cup \{z_n\}) = \frac{1}{\pi_a \cdot \mathbb{P}_a\left(\tau_{G\setminus G_n} < \tau_a^+\right)} .$$

Since $\bigcap_n \{\tau_{G\setminus G_n} < \tau_a^+\} = \{\tau_a^+ = \infty\}$ we have

$$\mathcal{R}_{\text{eff}}(a \leftrightarrow \infty) = \frac{1}{\pi_a \cdot \mathbb{P}_a(\tau_a^+ = \infty)} ,$$

with the convention that $1/0 = \infty$.

Definition 2.37 Let G be an infinite network. A function $\theta : E(G) \to \mathbb{R}$ is a **flow from a to ∞** if it is anti-symmetric and satisfies the node law on each vertex $v \neq a$.

The following follows easily from Theorem 2.28, we omit the proof.

Theorem 2.38 (Thomson's Principle for Infinite Networks) *Let G be an infinite network, then*

$$\mathcal{R}_{\text{eff}}(a \leftrightarrow \infty) = \inf\{\mathcal{E}(\theta) : \theta \text{ is a flow from } a \text{ to } \infty \text{ of strength } 1\}.$$

Corollary 2.39 *Let G be an infinite graph. The following are equivalent:*

1. *G is transient.*
2. *There exists a vertex $a \in V$ such that $\mathcal{R}_{\text{eff}}(a \leftrightarrow \infty) < \infty$. Hence all vertices satisfy this.*
3. *There exists a vertex $a \in V$ and a unit flow θ from a to ∞ with $\mathcal{E}(\theta) < \infty$. Hence all vertices satisfy this.*

We will now develop a useful method for bounding effective resistances from below. This will lead us to a popular sufficient criterion for recurrence in Corollary 2.43.

Definition 2.40 A **cutset** $\Gamma \subseteq E(G)$ separating a from z is a set of edges such that every path from a to z must use an edge from Γ.

Claim 2.41 Let θ be a flow from a to z in a finite network, and let Γ a cutset separating a from z. Then

$$\sum_{e \in \Gamma} |\theta(e)| \geq \|\theta\|.$$

Proof Denote by Z the set of vertices separated from a by Γ. Denote by G' the network where Z is identified to a single vertex x and all edges having both endpoints in Z are removed. Now, the restriction of θ to the edges of the new network is a flow from a to x. By Claim 2.17, we have $\sum_{y:y \sim x} \theta(yx) = \|\theta\|$. Also, all edges incident to x must be in Γ, since otherwise x is not separated from a by Γ. Therefore

$$\sum_{e \in \Gamma} |\theta(e)| \geq \sum_{y:y \sim x} \theta(yx) = \|\theta\|. \qquad \square$$

Theorem 2.42 (Nash-Williams Inequality) *Let $\{\Gamma_n\}$ be disjoint cutsets separating a from z in a finite network. Then*

$$\mathcal{R}_{\text{eff}}(a \leftrightarrow z) \geq \sum_n \left(\sum_{e \in \Gamma_n} c_e \right)^{-1}.$$

Proof Let θ be a flow from a to z with $\|\theta\| = 1$. From Cauchy-Schwarz, for each n we have

$$\left(\sum_{e \in \Gamma_n} \sqrt{r_e} \sqrt{c_e} |\theta(e)| \right)^2 \leq \sum_{e \in \Gamma_n} c_e \sum_{e \in \Gamma_n} r_e \theta(e)^2.$$

Also, since Γ_n is a cutset, the flow passing through Γ_n is at least $\|\theta\|$, by Claim 2.41. So

$$\left(\sum_{e \in \Gamma_n} \sqrt{r_e}\sqrt{c_e}|\theta(e)|\right)^2 \geq \|\theta\|^2 = 1.$$

Combining them, we get that

$$\sum_{e \in \Gamma_n} r_e\theta(e)^2 \geq \frac{1}{\sum_{e \in \Gamma_n} c_e}.$$

Summing over all n gives

$$\mathcal{E}(\theta) \geq \sum_{n}\sum_{e \in \Gamma_n} r_e\theta(e)^2 \geq \sum_{n}\left(\sum_{e \in \Gamma_n} c_e\right)^{-1}.$$

Applying Thomson's principle (Theorem 2.28) yields the result.   □

Consider now an infinite network $G = (V, E)$. We say that $\Gamma \subset E$ is a cutset separating a from ∞ if any infinite simple path from a must intersect Γ.

Corollary 2.43 *In any infinite network, if there exists a collection $\{\Gamma_n\}$ of disjoint cutsets separating a from ∞ such that*

$$\sum_{n}\left(\sum_{e \in \Gamma_n} c_e\right)^{-1} = \infty,$$

then the network is recurrent.

Example 2.44 ($\mathbb{Z}^2$ is Recurrent) Define Γ_n as the set of vertical edges $\{(x, y), (x, y+1)\}$ with $|x| \leq n$ and $\min\{|y|, |y+1|\} = n$ along with the horizontal edges $\{(x, y), (x+1, y)\}$ with $|y| \leq n$ and $\min\{|x|, |x+1|\} = n$, see Fig. 2.6. Then

Fig. 2.6 A part of $\mathbb{Z}^2$: the edges in $\{-1, 0, 1\}^2$ are drawn in bold. Γ_1 is dashed

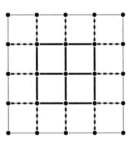

$\{\Gamma_n\}$ is a collection of disjoint cutsets separating 0 from ∞. Also, $|\Gamma_n| = 4(2n + 1)$ and therefore $\sum_n \left(\sum_{e \in \Gamma_n} c_e\right)^{-1} = \infty$. We deduce by Corollary 2.43 that $\mathbb{Z}^2$ is recurrent.

Remark 2.45 There are recurrent graphs for which there exists $M < \infty$ such that for every collection $\{\Gamma_n\}$ of disjoint cutsets, $\sum_n \left(\sum_{e \in \Gamma_n} c_e\right)^{-1} \leq M$. Therefore, the Nash-Williams inequality is not sharp. See Exercise 4 of this chapter.

2.6 Random Paths

We now present the method of random paths, which is one of the most useful methods for generating unit flows on a network and bounding their energy. In fact, it is possible to show that the electric flow can be represented by such a random path. Suppose G is a network with fixed vertices a, z and μ is a probability measure on the set of paths from a to z.

Claim 2.46 For a path γ sampled from μ, let

$$\theta_\gamma(\vec{e}) = (\text{\# of times } \vec{e} \text{ was traversed by } \gamma) - (\text{\# of times } \bar{e} \text{ was traversed by } \gamma),$$

where by $\vec{e}$ and $\bar{e}$ we mean the two orientations of an edge e of G. Set

$$\theta(\vec{e}) = \mathbb{E}\theta_\gamma(\vec{e}).$$

Then θ is a flow from a to z with $\|\theta\| = 1$.

Proof θ is antisymmetric since θ_γ is antisymmetric for every γ, and it satisfies the node law since θ_γ satisfies the node law. Similarly, the "strength" of θ_γ (i.e., $\sum_{x \sim a} \theta_\gamma(ax)$) is 1, hence $\|\theta\| = 1$.   □

An example of the use of this method is the following classical result.

Theorem 2.47 $\mathbb{Z}^3$ *is transient.*

Proof For $R > 0$ denote by $B_R = \{(x, y, z) : x^2 + y^2 + z^2 \leq R^2\}$ the ball of radius R in $\mathbb{R}^3$. Put $V_R = B_R \cap \mathbb{Z}^3$ and let ∂V_R be the external vertex boundary of V_R, that is, the set of vertices not in V_R which belong to an edge with an endpoint in V_R.

We construct a random path μ from the origin $\mathbf{0}$ to ∂V_R by choosing a uniform random point $\mathbf{p}$ in $\partial B_R = \{(x, y, z) : x^2 + y^2 + z^2 = R^2\}$, drawing a straight line between $\mathbf{0}$ and $\mathbf{p}$ in $\mathbb{R}^3$, considering the set of distance at most 10 in $\mathbb{R}^3$ from the line, and then choosing (in some arbitrary fashion) a path in $\mathbb{Z}^3$ which is contained inside this set. The non-optimal constant 10 was chosen in order to guarantee that such a discrete path exists for any point $\mathbf{p} \in \partial B_R$.

By Claim 2.46, the measure μ corresponds to a flow from $\mathbf{0}$ to ∂V_R. To estimate the energy of this flow, we note that if $\vec{e}$ is an edge at distance $r \leq R$ from the origin, then the probability that it is traversed by a path drawn by μ is $O(r^{-2})$. Furthermore, there are $O(r^2)$ such edges. Hence the energy of the flow is at most

$$\mathcal{E} = O\left(\sum_{r=1}^R r^2 \cdot (r^{-2})^2\right) \leq C,$$

for some constant $C < \infty$ which does not depend on R. By Claim 2.46 and Theorem 2.28 we learn that $\mathcal{R}_{\text{eff}}(0 \leftrightarrow \partial V_R) \leq C$ for all R, and so by Corollary 2.39 we deduce that $\mathbb{Z}^3$ is transient. □

2.7 Exercises

1. Let $G_z(a, x)$ be the Green's function, that is,

$$G_z(a, x) = \mathbb{E}_a\big[\#\text{visits to } x \text{ before visiting } z\big].$$

Show that the function $h(x) = G_z(a, x)/\pi(x)$ is a voltage.
2. Show that the effective resistance satisfies the triangle inequality. That is, for any three vertices x, y, z we have

$$\mathcal{R}_{\text{eff}}(x \leftrightarrow z) \leq \mathcal{R}_{\text{eff}}(x \leftrightarrow y) + \mathcal{R}_{\text{eff}}(y \leftrightarrow z). \tag{2.9}$$

3. Let a, z be two vertices of a finite network and let τ_a, τ_z be the first visit time to a and z, respectively, of the weighted random walk. Show that for any vertex x

$$\mathbf{P}_x(\tau_a < \tau_z) \leq \frac{\mathcal{R}_{\text{eff}}(x \leftrightarrow \{a, z\})}{\mathcal{R}_{\text{eff}}(x \leftrightarrow a)}.$$

4. Consider the following tree T. At height n it has 2^n vertices (the root is at height $n = 0$) and if $(v_1, \ldots, v_{2^n})$ are the vertices at level n we make it so that v_k has 1 child at level $n + 1$ and if $1 \leq k \leq 2^{n-1}$ and v_k has 3 children at level $n + 1$ for all other k.

   (a) Show that T is recurrent.
   (b) Show that for any disjoint edge cutsets Π_n we have that $\sum_n |\Pi_n|^{-1} < \infty$. (So, the Nash-Williams criterion for recurrence is not sharp)

5. (a) Let G be a finite planar graph with two distinct vertices $a \neq z$ such that a, z are on the outer face. Consider an embedding of G so that a is the left most point on the real axis and z is the right most point on the real axis. Split the outer face of G into two by adding the ray from a to $-\infty$ and the ray from z to $+\infty$. Consider the dual graph G^* of G and write a^* and z^* for the two vertices corresponding to the split outer face of G. Assume that all edge

resistances are 1. Show that

$$\mathcal{R}_{\text{eff}}(a \leftrightarrow z; G) = \frac{1}{\mathcal{R}_{\text{eff}}(a^* \leftrightarrow z^*; G^*)}.$$

(b) Show that the probability that a simple random walk on $\mathbb{Z}^2$ started at $(0, 0)$ has probability $1/2$ to visit $(0, 1)$ before returning to $(0, 0)$.

6. Let $G = (V, E)$ be a graph so that $V = \mathbb{Z}$ and the edge set $E = \cup_{k \geq 0} E_k$ where $E_0 = \{(i, i + 1) : i \in \mathbb{Z}\}$ and for $k > 0$

$$E_k = \left\{ \left(2^k (n - 1/2), 2^k (n + 1/2)\right) : n \in \mathbb{Z} \right\}.$$

Is G recurrent or transient?

Chapter 3
The Circle Packing Theorem

3.1 Planar Graphs, Maps and Embeddings

Definition 3.1 A graph $G = (V, E)$ is **planar** if it can be **properly drawn** in the plane, that is, if there exists a mapping sending the vertices to distinct points of $\mathbb{R}^2$ and edges to continuous curves between the corresponding vertices so that no two curves intersect, except at the vertices they share. We call such a mapping a **proper drawing** of G.

Remark 3.2 A single planar graph has infinitely many drawings. Intuitively, some may seem similar to one another, while others seem different. For example,

The following definition gives a precise sense to the above intuitive equivalence/non-equivalence of drawings.

Definition 3.3 A **planar map** is a graph endowed with a cyclic permutation of the edges incident to each vertex, such that there exists a proper drawing in which the clockwise order of the curves touching the image of a vertex respects that cyclic permutation.

The combinatorial structure of a planar map allows us to define faces directly (that is, without mentioning the drawing). Consider each edge of the graph as directed in both ways, and say that a directed edge $\vec{e}$ **precedes** $\vec{f}$ (or, equivalently,

A. Nachmias, *Planar Maps, Random Walks and Circle Packing*, Lecture Notes in Mathematics 2243, https://doi.org/10.1007/978-3-030-27968-4_3

Fig. 3.1 **Fig. 3.1** Examples for the edge precedence relation. (a) $\vec{e}$ precedes $\vec{f}$. (b) (x, y) precedes (y, x)

$\vec{f}$ **succeeds** $\vec{e}$), if there exist vertices v, x, y such that $\vec{e} = (x, v)$, $\vec{f} = (v, y)$, and y is the successor of x in the cyclic permutation σ_v; see Fig. 3.1.

We say that $\vec{e}, \vec{f}$ **belong to the same face** if there exists a finite directed path $\vec{e}_1, \ldots, \vec{e}_m$ in the graph with $\vec{e}_i$ preceding $\vec{e}_{i+1}$ for $i = 1, \ldots, m - 1$ and such that either $\vec{e} = \vec{e}_1$ and $\vec{f} = \vec{e}_m$, or $\vec{f} = \vec{e}_1$ and $\vec{e} = \vec{e}_m$. This is readily seen to be an equivalence relation and we call each equivalence class a **face**. Even though a face is a set of directed edges, we frequently ignore the orientations and consider a face as the set of corresponding undirected edges. Each (undirected) edge is henceforth incident to either one or two faces.

When the map is finite an equivalent definition of a face is the set of edges that bound a connected component of the complement of the drawing, that is, of $\mathbb{R}^2$ minus the images of the vertices and edges. This definition is not suitable for infinite planar maps since there may be a complicated set of accumulation points. Given a proper drawing of a finite planar map, there is a unique unbounded connected component of the complement of the drawing; the edges that bound it are called the **outer face** and all other faces are called **inner faces** . However, for any face in a finite map there is a drawing so that this face bounds the unique unbounded connected component, and because of this we shall henceforth refer to the outer face as an arbitrarily chosen face of the map.

We will use the following classical formula.

Theorem 3.4 (Euler's Formula)   *Suppose G is a planar graph with n vertices, m edges and f faces. Then*

$$n - m + f = 2.$$

We now state the main theorem we will discuss and use throughout this course. Its proof is presented in the next section.

Theorem 3.5 (The Circle Packing Theorem [51])   *Given any finite simple planar map $G = (V, E)$, $V = \{v_1, \ldots, v_n\}$, there exist n circles in $\mathbb{R}^2$, $C_1, \ldots, C_n$, with disjoint interiors, such that C_i is tangent to C_j if and only if $\{i, j\} \in E$.*

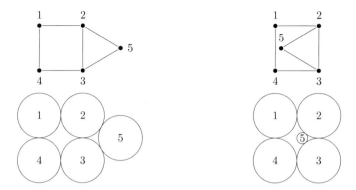

Fig. 3.2   Two distinct planar maps (of the same graph) with corresponding circle packings

Furthermore, for every vertex v_i, the clockwise order of the circles tangent to C_i agrees with the cyclic permutation of v_i's neighbors in the map.

Figure 3.2 gives examples for embeddings of maps which respect the cyclic orderings of neighbors, as guaranteed to exist according to the theorem.

First note that it suffices to prove the theorem for **triangulations**, that is, simple planar maps in which every face has precisely three edges. Indeed, in any planar map we may add a single vertex inside each face and connect it to all vertices bounding that face. The obtained map is a triangulation, and after applying the circle packing theorem for triangulations, we may remove the circles corresponding to the added vertices, obtaining a circle packing of the original map which respects its cyclic permutations. This is depicted in Fig. 3.3.

Thus, it suffices to prove Theorem 3.5 for finite triangulations. In this case an important uniqueness statement also holds.

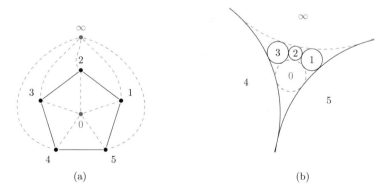

Fig. 3.3   Circle packing of a triangulation of a planar map. (**a**) A planar map and a triangulation. (**b**) A circle packing of the triangulation

Theorem 3.6 *Let $G = (V, E)$ be a finite triangulation on vertex set $V = \{v_1, \ldots, v_n\}$ and assume that $\{v_1, v_2, v_3\}$ form a face. Then for any three positive numbers ρ_1, ρ_2, ρ_3, there exists a circle packing $C_1, \ldots, C_n$ as in Theorem 3.5 with the additional property that C_1, C_2, C_3 are mutually tangent, form the outer face, and have radii ρ_1, ρ_2, ρ_3, respectively. Furthermore, this circle packing is unique, up to translations and rotations of the plane.*

3.2   Proof of the Circle Packing Theorem

We prove Theorem 3.6 which implies Theorem 3.5 as explained above. Therefore we assume from now on that our map is a triangulation. Denote by n, m and f the number of vertices, edges and faces of the map respectively, and observe that $3f = 2m$ since each edge is counted in exactly two faces, and each face is bounded by exactly three edges. Therefore, by Euler's formula (Theorem 3.4), we have that

$$2 = n - m + f = n - \frac{3}{2}f + f = n - \frac{1}{2}f,$$

thus

$$f = 2n - 4. \tag{3.1}$$

We assume the vertex set is $\{v_1, \ldots, v_n\}$, that $\{v_1, v_2, v_3\}$ is the outer face and that ρ_1, ρ_2, ρ_3 are three positive numbers that will be the radii of the outer circles C_1, C_2, C_3 eventually. Denote by F° the set of inner faces of the map, and for a subset of vertices A let $F(A)$ be the set of inner faces with at least one vertex in A. We write $F(v)$ when we mean $F(\{v\})$.

Given a vector $\mathbf{r} = (r_1, \ldots, r_n) \in (0, \infty)^n$, an inner face $f \in F^\circ$ bounded by the vertices v_i, v_j, v_k, and a distinguished vertex v_j, we associate a number $\alpha_f^\mathbf{r}(v_j) = \angle v_i v_j v_k \in (0, \pi)$ which is the angle of v_j in the triangle $\triangle v_i v_j v_k$ created by connecting the centers of three mutually tangent circles C_i, C_j, C_k of radii r_i, r_j and r_k (that is, in a triangle with side lengths $r_i + r_j$, $r_j + r_k$ and $r_k + r_i$). This number can be calculated using the cosine formula

$$\cos(\angle v_i v_j v_k) = 1 - \frac{2r_i r_k}{(r_i + r_j)(r_j + r_k)},$$

however, we will not use this formula directly. For every $j \in \{1, \ldots, n\}$ we define

$$\sigma_\mathbf{r}(v_j) = \sum_{f \in F(v_j)} \alpha_f^\mathbf{r}(v_j)$$

to be the *sum of angles* at v_i with respect to $\mathbf{r}$. Let $\theta_1, \theta_2, \theta_3$ be the angles formed at the centers of three mutually tangent circles C_1, C_2, C_3 of radii ρ_1, ρ_2, ρ_3. Equivalently, these are the angles of a triangle with edge lengths $r_1 + r_2$, $r_2 + r_3$ and $r_1 + r_3$. If the vector $\mathbf{r}$ was the vector of radii of a circle packing of the map satisfying Theorem 3.6, then it would hold that

$$\sigma_{\mathbf{r}}(v_i) = \begin{cases} \theta_i & i \in \{1, 2, 3\}, \\ 2\pi & \text{otherwise}, \end{cases} \tag{3.2}$$

and additionally $(r_1, r_2, r_3) = (\rho_1, \rho_2, \rho_3)$. The proof is split into three parts:

1. Show that there exists a vector $\mathbf{r} \in (0, \infty)^n$ satisfying (3.2);
2. Given such $\mathbf{r}$, show that a circle packing with these radii exists and that (r_1, r_2, r_3) is a positive multiple of (ρ_1, ρ_2, ρ_3); furthermore, this circle packing is unique up to translations and rotations.
3. Show that $\mathbf{r}$ is unique up to scaling all entries by a constant factor.

Proof of Theorem 3.6, Step 1: Finding the Radii Vector $\mathbf{r}$

Observation 3.7 *For every* $\mathbf{r}$,

$$\sum_{i=1}^{n} \sigma_{\mathbf{r}}(v_i) = |F^\circ| \pi = (2n - 5)\pi.$$

Proof Follows immediately since each inner face f bounded by the vertices v_i, v_j, v_k contributes the three angles $\alpha_f^{\mathbf{r}}(v_i)$, $\alpha_f^{\mathbf{r}}(v_j)$ and $\alpha_f^{\mathbf{r}}(v_k)$ which sum to π. By (3.1), there are $2n - 5$ inner faces. □

We now set

$$\delta_{\mathbf{r}}(v_i) = \begin{cases} \sigma_{\mathbf{r}}(v_i) - \theta_i & j \in \{1, 2, 3\}, \\ \sigma_{\mathbf{r}}(v_j) - 2\pi & \text{otherwise}. \end{cases} \tag{3.3}$$

Using this notation, our goal is to find $\mathbf{r}$ for which $\delta_{\mathbf{r}} \equiv 0$. It follows from Observation 3.7 that for every $\mathbf{r}$,

$$\sum_{i=1}^{n} \delta_{\mathbf{r}}(v_i) = \sum_{i=1}^{n} \sigma_{\mathbf{r}}(v_i) - \theta_1 - \theta_2 - \theta_3 - (n - 3) \cdot 2\pi = 0. \tag{3.4}$$

We define

$$\mathcal{E}_{\mathbf{r}} = \sum_{i=1}^{n} \delta_{\mathbf{r}}(v_i)^2.$$

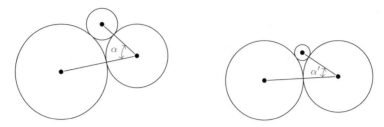

Fig. 3.4   When the radius of a circle corresponding to a vertex increases, while the radii of the circles corresponding to its two neighbors in a given face decrease, the vertex's angle in the corresponding triangle decreases (see Observation 3.8)

We would like to find $\mathbf{r}$ for which $\mathcal{E}_{\mathbf{r}} = 0$. We will use the following geometric observation; see Fig. 3.4.

Observation 3.8 *Let* $\mathbf{r} = (r_1, \ldots, r_n)$ *and* $\mathbf{r}' = (r_1', \ldots, r_n')$, *and let* $f \in F^\circ$ *be bounded by* v_i, v_j, v_k.

- *If* $r_i' \leq r_i$, $r_k' \leq r_k$ *and* $r_j' \geq r_j$, *then* $\alpha_f^{\mathbf{r}'}(v_j) \leq \alpha_f^{\mathbf{r}}(v_j)$.
- *If* $r_i' \geq r_i$, $r_k' \geq r_k$ *and* $r_j' \leq r_j$, *then* $\alpha_f^{\mathbf{r}'}(v_j) \geq \alpha_f^{\mathbf{r}}(v_j)$.
- $\alpha_f^{\mathbf{r}}(v_j)$ *is continuous in* $\mathbf{r}$.

Proof A proof using the cosine formula is routine and is omitted.   □

We now define an iterative algorithm, whose input and output are both vectors of radii normalized to have ℓ_1 norm 1. We start with the vector $\mathbf{r}^{(0)} = \left(\frac{1}{n}, \ldots, \frac{1}{n}\right)$, and given $\mathbf{r} = \mathbf{r}^{(t)}$ we construct $\mathbf{r}' = \mathbf{r}^{(t+1)}$. Write $\delta = \delta_{\mathbf{r}}$ and $\delta' = \delta_{\mathbf{r}'}$, and similarly $\mathcal{E} = \mathcal{E}_{\mathbf{r}}$ and $\mathcal{E}' = \mathcal{E}_{\mathbf{r}'}$. We begin by ordering the set of reals $\{\delta(v_i) \mid 1 \leq i \leq n\}$. If $\delta \equiv 0$ we are done; otherwise, we may choose $s \in \mathbb{R}$ such that the set $S = \{v \mid \delta(v) > s\} \neq \varnothing$ and its complement $V \setminus S$ are non-empty and such that the gap

$$\mathrm{gap}_\delta(S) := \min_{v \in S} \delta(v) - \max_{v \notin S} \delta(v) > 0$$

is maximal over all such s. See Fig. 3.5 for illustration.

Fig. 3.5   Left: finding the maximum gap between two consecutive values of δ, and splitting the set of values into S and its complement. Right: moving from $\mathbf{r}$ to $\mathbf{r}'$ closes the gap between S and $V \setminus S$

Once we choose S, a step of the algorithm consists of two steps:

1. For some $\lambda \in (0, 1)$ to be chosen later, we set

$$(\mathbf{r}_\lambda)_i = \begin{cases} r_i & v_i \in S, \\ \lambda r_i & v_i \notin S. \end{cases}$$

2. We normalize $\mathbf{r}_\lambda$ so that the sum of entries is 1, letting $\bar{\mathbf{r}}_\lambda$ be the normalized vector. Note that this step does not change the vector δ.

We will choose an appropriate λ that will decrease all values of $\delta(v)$ for $v \in S$, increase all values of $\delta(v)$ for $v \notin S$, and will close the gap. This is made formal in the following two claims.

Claim 3.9 For every $\lambda \in (0, 1)$, setting $\mathbf{r}' = \bar{\mathbf{r}}_\lambda$, we have that $\delta'(v) \leq \delta(v)$ for any $v \in S$, and $\delta'(v) \geq \delta(v)$ for any $v \notin S$.

Claim 3.10 There exists $\lambda \in (0, 1)$ such that setting $\mathbf{r}' = \bar{\mathbf{r}}_\lambda$ gives that $\mathrm{gap}_{\delta'}(S) = 0$.

Proof of Claim 3.9 Consider $v_j \notin S$ and an inner face v_i, v_j, v_k.

Case I $v_i, v_k \notin S$. In this case, the radii of C_i, C_j, C_k are all multiplied by the same number λ, so $\alpha_f^{\mathbf{r}'}(v_j) = \alpha_f^{\mathbf{r}}(v_j)$.

Case II $v_i, v_k \in S$. In this case, the radii of C_i, C_k remain unchanged and the radius of C_j decreases, thus by Observation 3.8, $\alpha_f^{\mathbf{r}'}(v_j) \geq \alpha_f^{\mathbf{r}}(v_j)$.

Case III $v_i \notin S$, $v_k \in S$. In this case the radii of C_i, C_j are multiplied by λ and the radius of C_k is unchanged. The angles of $\triangle v_i v_j v_k$ remain unchanged if we multiply all radii by λ^{-1}, thus we could just as easily have left C_i, C_j unchanged and increased the radius of C_k. By Observation 3.8, we get that $\alpha_f^{\mathbf{r}'}(v_j) \geq \alpha_f^{\mathbf{r}}(v_j)$.

It follows that $\delta'(v) \geq \delta(v)$ for any $v \notin S$. An identical argument shows that $\delta'(v) \leq \delta(v)$ for all $v \in S$. □

In order to prove Claim 3.10, we present another claim.

Claim 3.11

$$\lim_{\lambda \searrow 0} \sum_{v \notin S} \delta_{\mathbf{r}_\lambda}(v) > 0.$$

Proof of Claim 3.10 Using Claim 3.11 The function $\lambda \mapsto \mathrm{gap}_{\delta_{\mathbf{r}_\lambda}}(S)$ is continuous on $(0, 1]$ by the third bullet of Observation 3.8, and its value at $\lambda = 1$ is $\mathrm{gap}_\delta(S) > 0$. Claim 3.11 says that if $\mu > 0$ is small enough, then

$$\sum_{v \notin S} \delta_{\mathbf{r}_\mu}(v) > 0,$$

from which it follows that $\max_{v \notin S} \delta_{\mathbf{r}_\mu}(v) > 0$. By (3.4), we also have

$$\sum_{v \in S} \delta_{\mathbf{r}_\mu}(v) < 0,$$

meaning that $\min_{v \in S} \delta_{\mathbf{r}_\mu}(v) < 0$ and therefore $\mathrm{gap}_{\delta_{\mathbf{r}_\mu}}(S) < 0$. By continuity, there exists $\lambda \in (\mu, 1)$ such that $\mathrm{gap}_{\delta_{\mathbf{r}_\lambda}}(S) = 0$.   □

Proof of Claim 3.11 We first show that for each face $f \in F(V \setminus S)$ bounded by v_i, v_j, v_k, the sum of angles at the vertices belonging to $V \setminus S$ converges to π as $\lambda \searrow 0$. We show this by the following case analysis. The statements in cases II and III can be justified by drawing a picture or appealing to the cosine formula.

Case I   If $v_i, v_j, v_k \notin S$ then since the face is a triangle, $\alpha_f^{\mathbf{r}_\lambda}(v_i) + \alpha_f^{\mathbf{r}_\lambda}(v_j) + \alpha_f^{\mathbf{r}_\lambda}(v_k) = \pi$ for all $\lambda \in (0, 1)$.

Case II   If $v_i, v_j \notin S$ but $v_k \in S$ then $\lim_{\lambda \searrow 0} \alpha_f^{\mathbf{r}_\lambda}(v_k) = 0$, hence $\lim_{\lambda \searrow 0} \alpha_f^{\mathbf{r}_\lambda}(v_i) + \alpha_f^{\mathbf{r}_\lambda}(v_j) = \pi$.

Case III   If $v_i \notin S$ but $v_j, v_k \in S$ then $\lim_{\lambda \searrow 0} \alpha_f^{\mathbf{r}_\lambda}(v_j) + \alpha_f^{\mathbf{r}_\lambda}(v_k) = 0$, hence $\lim_{\lambda \searrow 0} \alpha_f^{\mathbf{r}_\lambda}(v_i) = \pi$.

It follows that

$$\lim_{\lambda \searrow 0} \sum_{v \notin S} \sigma_{\mathbf{r}_\lambda}(v) = |F(V \setminus S)|\pi. \tag{3.5}$$

For convenience, set

$$\theta(v_i) = \begin{cases} \theta_i & 1 \le i \le 3, \\ 2\pi & \text{otherwise}, \end{cases}$$

so that $\delta_{\mathbf{r}}(v) = \sigma_{\mathbf{r}}(v) - \theta(v)$ for all $v \in V$. Then

$$\lim_{\lambda \searrow 0} \sum_{v \notin S} \delta_{\mathbf{r}_\lambda}(v) = |F(V \setminus S)|\pi - \sum_{v \notin S} \theta(v). \tag{3.6}$$

Let $\bar{F} = F^\circ \setminus F(V \setminus S)$, so every face in $\bar{F}$ contains only vertices of S. We will show that

$$|\bar{F}|\pi < \sum_{v \in S} \theta(v). \tag{3.7}$$

If (3.7) holds, then we can add the negative quantity $|\bar{F}|\pi - \sum_{v \in S} \theta(v)$ to the right side of (3.6), obtaining $|F^\circ|\pi - \sum_{v \in V} \theta(v) = (2n-5)\pi - (2n-5)\pi = 0$. It follows that (3.6) is strictly positive, proving the claim. Thus it suffices to show (3.7).

In the rest of the proof, we fix an embedding of G in the plane with (v_1, v_2, v_3) as the outer face. Let $G[S]$ be the subgraph of G induced by S. Partition S into equivalence classes, $S = S_1 \cup \cdots \cup S_k$, where two vertices are equivalent if they are in the same connected component of $G[S]$. Then $G[S] = G[S_1] \cup \cdots \cup G[S_k]$. Let $\bar{F}_j$ be the set of faces in $\bar{F}$ that appear as faces of $G[S_j]$, so that we have the disjoint union $\bar{F} = \bar{F}_1 \cup \cdots \cup \bar{F}_k$.

Since S is nonempty, it is enough to show that for all $1 \leq j \leq k$,

$$|\bar{F}_j|\pi < \sum_{v \in S_j} \theta(v). \tag{3.8}$$

Let m_j and f_j denote the number of edges and faces, respectively, of $G[S_j]$. Observe that $|\bar{F}_j| \leq f_j - 1$. If $|\bar{F}_j| = 0$, then (3.8) is trivial. If $|\bar{F}_j| \geq 1$, then $G[S_j]$ has at least one inner face, and since it is a simple graph, every face must have degree at least 3. (The degree of a face is the number of directed edges that make up its boundary.) Because the sum of the degrees of all the faces equals twice the number of edges, we have $2m_j \geq 3f_j$. Euler's formula now gives

$$|S_j| + f_j - 2 = m_j \geq \frac{3}{2} f_j,$$

and hence $f_j \leq 2|S_j| - 4$. Thus, the left side of (3.8) satisfies

$$|\bar{F}_j|\pi \leq (2|S_j| - 5)\pi.$$

If S_j contains all of v_1, v_2, v_3, then the right side of (3.8) is

$$\theta_1 + \theta_2 + \theta_3 + (|S_j| - 3) \cdot 2\pi = (2|S_j| - 5)\pi.$$

Otherwise, at least one of the θ_i is replaced by 2π and so the right side of (3.8) is strictly greater than the left side. In fact, (3.8) holds except when $v_1, v_2, v_3 \in S_j$ and $|\bar{F}_j| = f_j - 1 = 2|S_j| - 5$. We now show that this situation cannot occur.

The equality $|\bar{F}_j| = f_j - 1$ means that every inner face of $G[S_j]$ is an element of $\bar{F}_j$ and therefore a face of G. Since $v_1, v_2, v_3 \in S_j$, the outer face of $G[S_j]$ is (v_1, v_2, v_3), which is the same as the outer face of G. So, every face of $G[S_j]$ is also a face of G. But this is impossible: if we choose any $v \in V \setminus S$, then v must lie in some face of $G[S_j]$, which then cannot be a face of G. Therefore, it cannot be true that $v_1, v_2, v_3 \in S_j$ and also $|\bar{F}_j| = f_j - 1$, so we conclude that (3.8) always holds. □

We now analyse the algorithm. Let $\lambda \in (0, 1)$ be the one guaranteed by Claim 3.10, and set $\mathbf{r}' = \bar{\mathbf{r}}_\lambda$.

Claim 3.12 $\mathcal{E}' \leq \mathcal{E}\left(1 - \frac{1}{2n^3}\right).$

Proof As depicted in Fig. 3.5, define

$$t = \min_{v \in S} \delta'(v) = \max_{v \notin S} \delta'(v).$$

By (3.4) we have that $\sum_{i=1}^{n} \delta(v_i) = \sum_{i=1}^{n} \delta'(v_i) = 0$, hence

$$\mathcal{E} - \mathcal{E}' = \sum_{i=1}^{n} \delta(v_i)^2 - \sum_{i=1}^{n} \delta'(v_i)^2 = \sum_{i=1}^{n} (\delta(v_i) - \delta'(v_i))^2 + 2\sum_{i=1}^{n} (t - \delta'(v_i))(\delta'(v_i) - \delta(v_i)).$$

If $v \in S$, then $t \leq \delta'(v) \leq \delta(v)$ and if $v \notin S$, then $t \geq \delta'(v) \geq \delta(v)$. Thus, in both cases $(t - \delta'(v))(\delta'(v) - \delta(v)) \geq 0$. Taking $u \in S$ and $v \notin S$ with $\delta'(u) = \delta'(v) = t$, we have that

$$\mathcal{E} - \mathcal{E}' \geq (\delta(u) - t)^2 + (\delta(v) - t)^2 \geq \frac{(\delta(u) - \delta(v))^2}{2} \geq \frac{\text{gap}_\delta(S)^2}{2}.$$

Since $\text{gap}_\delta(S)$ was chosen to be the maximal gap we may bound,

$$\text{gap}_\delta(S) \geq \frac{1}{n}\left(\max_{v \in V} \delta(v) - \min_{v \in V} \delta(v)\right).$$

For every $v \in V$,

$$\max_{w \in V} \delta(w) - \min_{w \in V} \delta(w) \geq |\delta(v)|,$$

and thus

$$n\left(\max_{v \in V} \delta(v) - \min_{v \in V} \delta(v)\right)^2 \geq \sum_{i=1}^{n} \delta(v_i)^2 = \mathcal{E}.$$

Hence

$$\mathcal{E} - \mathcal{E}' \geq \frac{1}{2n^2}\left(\max_{v \in V} \delta(v) - \min_{v \in V} \delta(v)\right)^2 \geq \frac{1}{2n^2} \cdot \frac{\mathcal{E}}{n},$$

and we conclude that

$$\mathcal{E}' \leq \mathcal{E}\left(1 - \frac{1}{2n^3}\right). \qquad \square$$

Write $\mathcal{E}^{(t)} = \mathcal{E}_{\mathbf{r}^{(t)}}$. By iterating the described algorithm, we obtain from Claim 3.12 that

$$\mathcal{E}^{(t)} \leq \mathcal{E}^{(0)} \left(1 - \frac{1}{2n^3}\right)^t \longrightarrow 0 \qquad \text{as } t \to \infty.$$

By our normalization $\left\|\mathbf{r}^{(t)}\right\|_{\ell_1} = 1$. Thus, by compactness, there exists a subsequence $\{t_k\}$ and a vector $\mathbf{r}^\infty$ such that $\mathbf{r}^{(t_k)} \to \mathbf{r}^\infty$ as $k \to \infty$. From continuity of $\mathcal{E}$ we have that $\mathcal{E}(\mathbf{r}^\infty) = 0$, meaning that (3.2) is satisfied. For $\mathbf{r}^\infty$ to be feasible as a vector of radii, we also have to argue that it is positive (the fact that no coordinates are ∞ follows since $\|\mathbf{r}^\infty\|_{\ell_1} = 1$).

Claim 3.13 $\mathbf{r}_i^\infty > 0$ for every i.

Proof Let $S = \{v_i \in V : \mathbf{r}_i^\infty > 0\}$. Because of the normalization of $\mathbf{r}$, we know that S is nonempty. Assume for contradiction that $S \subsetneq V$. We repeat the exact same argument used in the proof of Claim 3.11 showing first by case analysis that

$$\lim_{t \to \infty} \sum_{v \notin S} \sigma_{\mathbf{r}^{(t)}}(v) = |F(V \setminus S)|\pi$$

and then deducing that

$$\lim_{t \to \infty} \sum_{v \notin S} \delta_{\mathbf{r}^{(t)}}(v) > 0.$$

This contradicts that $\lim_{t \to \infty} \mathcal{E}^{(t)} = 0$, so we conclude that $S = V$. □

Proof of Theorem 3.6, Step 2: Drawing the Circle Packing Described by $\mathbf{r}^\infty$

Given the vector of radii $\mathbf{r}^\infty$ satisfying (3.2), we now show that the corresponding circle packing can be drawn uniquely up to translations and rotations. In fact, we provide a slightly more general statement which is due to Ori Gurel-Gurevich and Ohad Feldheim [personal communications, 2018].

Let $G = (V, E)$ be a finite planar triangulation on vertex set $\{v_1, \ldots, v_n\}$ and assume that $\{v_1, v_2, v_3\}$ is the outer face. A vector of positive real numbers $\ell = \{\ell_e\}_{e \in E}$ indexed by the edge set E is called *feasible* if for any face enclosed by edges e_1, e_2, e_3, the lengths $\ell_{e_1}, \ell_{e_2}, \ell_{e_3}$ can be made to form a triangle. In other words, these lengths satisfy three triangle inequalities,

$$\ell_{e_i} + \ell_{e_j} > \ell_{e_k} \qquad \{i, j, k\} = \{1, 2, 3\}.$$

Given a feasible edge length vector ℓ we may again use the cosine formula to compute, for each face f, the angle at a vertex of the triangle formed by the three corresponding edge lengths. We denote these angles, as before, by $\alpha_f^\ell(v)$ where v is a vertex of f. Similarly, we define

$$\sigma_\ell(v) = \sum_{f \in F(v)} \alpha_f^\ell(v)$$

to be the sum of angles at a vertex v.

Theorem 3.14 *Let G be a finite triangulation and ℓ a feasible vector of edge lengths. Assume that $\sigma_\ell(v) = 2\pi$ for any internal vertex v. Then there is a drawing of G in the plane so that each edge e is drawn as a straight line segment of length ℓ_e and no two edges cross. Furthermore, this drawing is unique up to translations and rotations.*

It is easy to use the theorem above to draw the circle packing given the radii vector $\mathbf{r}^\infty$ satisfying (3.2). Indeed, given $\mathbf{r}^\infty$ we set ℓ by putting $\ell_e = \mathbf{r}_i^\infty + \mathbf{r}_j^\infty$ for any edge $e = \{v_i, v_j\}$ of the graph. Condition (3.2) implies that ℓ is feasible. We now apply Theorem 3.14 and obtain the guaranteed drawing and draw a circle C_i of radii $\mathbf{r}_i^\infty$ around v_i for all i. Theorem 3.14 guarantees that for any edge $\{v_i, v_k\}$ the distance between v_i, v_j is precisely $\mathbf{r}_i^\infty + \mathbf{r}_j^\infty$ and thus C_i and C_j are tangent. Conversely, assume that v_i, v_j do not form an edge. To each vertex v let A_v be the union of triangles touching v, each triangle is the space bounded by a face touching v in the drawing of Theorem 3.14. Since G is a triangulation and v_i and v_j are not adjacent we learn that A_{v_i} and A_{v_j} have disjoint interiors. Furthermore, $C_i \subset \text{Int}(A_{v_i})$ since the straight lines emanating from v_i have length larger than $\mathbf{r}_i^\infty$. By the same token $C_j \subset \text{Int}(A_{v_j})$ and we conclude that C_i and C_j are not tangent.

Lastly, we note that by (3.2) the outer boundary of the polygon we drew is a triangle with angles $\theta_1, \theta_2, \theta_3$ and hence (r_1, r_2, r_3) is a positive multiple of (ρ_1, ρ_2, ρ_3). Step 2 of the proof of Theorem 3.5 is now concluded.

Proof of Theorem 3.14 We prove this by induction on the number of vertices n. The base case $n = 3$ is trivial since the feasibility of ℓ guarantees that the edge lengths of the three edges of the outer face can form a triangle. Any two triangles with the same edge lengths can be rotated and translated to be identical, so the uniqueness statement holds for $n = 3$.

Assume now that $n > 3$ so that there exists an internal vertex v. Denote by $v_1, \ldots, v_m$ the neighbors of v ordered clockwise. We begin by placing v at the origin and drawing all the faces to which v belongs, see Fig. 3.6, left. That is, we draw the edge $\{v, v_1\}$ as a straight line interval of length $\ell_{\{v,v_1\}}$ on the positive x-axis emanating from the origin and proceed iteratively: for each $1 < i \le m$ we draw the edge $\{v, v_i\}$ as a straight line interval of length $\ell_{\{v,v_i\}}$ emanating from the origin (v) at a clockwise angle of $\alpha_f^\ell(v)$ from the previous drawn line segment of $\{v, v_{i-1}\}$, where $f = \{v, v_{i-1}, v_i\}$. This determines the location of $v_1, \ldots, v_m$ in the plane and allows us to "complete" the triangles by drawing the straight line

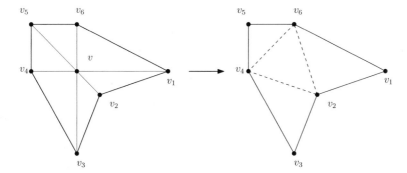

Fig. 3.6 On the left, we first draw the polygon surrounding v. On the right, we then erase v and the edge emanating from it, replacing it with diagonals that triangulate the polygon while recording the lengths of the diagonals in ℓ'. The latter is the input to the induction hypothesis

segments connecting v_i to v_{i+1}, each of length $\ell_{\{v_i, v_{i+1}\}}$ where $1 \leq i \leq m$ (where $v_{m+1} = v_1$). Denote these edges by $e_1, \ldots, e_m$.

Since $\sigma_\ell(v) = 2\pi$ we learn that these m triangles have disjoint interiors and that the edges $e_1, \ldots, e_m$ form a closed polygon containing the origin in its interior. It is a classical fact [64] that every closed polygon can be triangulated by drawing some diagonals as straight line segments in the interior of the polygon. We fix such a choice of diagonals and use it to form a new graph G' on $n - 1$ vertices and $|E(G)| - 3$ edges by erasing v and the m edges emanating from it and adding the new $m - 3$ edges corresponding to the diagonals we added. Furthermore, we generate a new edge length vector ℓ' corresponding to G' by assigning the new edges lengths corresponding to the Euclidean length of the drawn diagonals and leaving the other edge lengths unchanged. See Fig. 3.6, right.

It is clear that ℓ' is feasible and that the angle sum at each internal vertex of G' is 2π. Therefore we may apply the induction hypothesis and draw the graph G' according to the edge lengths ℓ'. This drawing is unique up to translations and rotations by induction. Note that in this drawing of G', the polygon corresponding to $e_1, \ldots, e_m$ must be the exact same polygon as before, up to translations and rotations, since it has the same edge lengths and the same angles between its edges. Since it is the same polygon, we can now erase the diagonals in this drawing and place a new vertex in the same relative location where we drew v previously, along with the straight line segments connecting it to $v_1, \ldots, v_m$. Thus we have obtained the desired drawing of G. The uniqueness up to translations and rotations of this drawing follows from the uniqueness of the drawing of G' and the fact that the location of v is uniquely determined in that drawing. $\square$

Proof of Theorem 3.6, Step 3: Uniqueness

Theorem 3.15 (Uniqueness of Circle Packing) *Given a simple finite triangulation with outer face v_1, v_2, v_3 and three radii ρ_1, ρ_2, ρ_3, the circle packing with $C_{v_1}, C_{v_2}, C_{v_3}$ having radii ρ_1, ρ_2, ρ_3 is unique up to translations and rotations.*

Proof We have already seen in step 2 that given the radii vector $\mathbf{r}$ the drawing we obtain is unique up to translations and rotations. Thus, we only need to show the uniqueness of $\mathbf{r}$ given ρ_1, ρ_2, ρ_3.

To that aim, suppose that $\mathbf{r}^a$ and $\mathbf{r}^b$ are two vectors satisfying (3.2). Since the outer face in both vectors correspond to a triangle of angles $\theta_1, \theta_2, \theta_3$ we may rescale so that $\mathbf{r}_i^a = \mathbf{r}_i^b = \rho_i$ for $i = 1, 2, 3$. After this rescaling, assume by contradiction that $\mathbf{r}^a \neq \mathbf{r}^b$ and let v be the interior vertex which maximizes $\mathbf{r}_v^a/\mathbf{r}_v^b$. We can assume without loss of generality that this quantity is strictly larger than 1, as otherwise we can swap $\mathbf{r}^a$ and $\mathbf{r}^b$.

Now we claim that for each $f = (v, u_1, u_2) \in F(v)$, we have $\alpha_f^{\mathbf{r}^a}(v) \leq \alpha_f^{\mathbf{r}^b}(v)$, with equality if and only if the ratios $\mathbf{r}_{u_i}^a/\mathbf{r}_{u_i}^b$, for $i = 1, 2$, are both equal to $\mathbf{r}_v^a/\mathbf{r}_v^b$. This is a direct consequence of Observation 3.8. Indeed, scale all the radii in $\mathbf{r}^b$ by a factor of $\mathbf{r}_v^a/\mathbf{r}_v^b$ to get a new vector $\mathbf{r}'$ such that $\mathbf{r}_v^a = \mathbf{r}_v'$ and $\mathbf{r}_u^a \leq \mathbf{r}_u'$ for all $u \neq v$. The second bullet point in Observation 3.8 implies that $\alpha_f^{\mathbf{r}^a}(v) \leq \alpha_f^{\mathbf{r}'}(v) = \alpha_f^{\mathbf{r}^b}(v)$. As well, if either $\mathbf{r}_{u_1}^a < \mathbf{r}_{u_1}'$ or $\mathbf{r}_{u_2}^a < \mathbf{r}_{u_2}'$, then the cosine formula yields the strict inequality $\alpha_f^{\mathbf{r}^a}(v) < \alpha_f^{\mathbf{r}'}(v)$. Thus, $\alpha_f^{\mathbf{r}^a}(v) = \alpha_f^{\mathbf{r}^b}(v)$ only if $\mathbf{r}_{u_i}^a/\mathbf{r}_{u_i}^b = \mathbf{r}_v^a/\mathbf{r}_v^b$ for $i = 1, 2$.

Now, since $\alpha_f^{\mathbf{r}^a}(v) \leq \alpha_f^{\mathbf{r}^b}(v)$ for each $f \in F(v)$, while $\sigma_{\mathbf{r}^a}(v) = \sigma_{\mathbf{r}^b}(v) = 2\pi$, the equality $\alpha_f^{\mathbf{r}^a}(v) = \alpha_f^{\mathbf{r}^b}(v)$ must hold for each f. Therefore, each neighbor u of v satisfies $\mathbf{r}_u^a/\mathbf{r}_u^b = \mathbf{r}_v^a/\mathbf{r}_v^b$. Because the graph is connected, the ratio $\mathbf{r}_u^a/\mathbf{r}_u^b$ must be constant for all vertices $u \in V(G)$. But this contradicts that $\mathbf{r}_v^a/\mathbf{r}_v^b > 1$ while $\mathbf{r}_{v_i}^a/\mathbf{r}_{v_i}^b = 1$ for $i = 1, 2, 3$. We conclude that $\mathbf{r}^a = \mathbf{r}^b$. $\qquad\square$

Chapter 4
Parabolic and Hyperbolic Packings

4.1 Infinite Planar Maps

In this chapter we discuss countably infinite connected simple graphs that are **locally finite**, that is, the vertex degrees are finite. In a similar fashion to the previous chapter, an infinite planar graph is a connected infinite graph such that there exists a drawing of it in the plane. We recall that a *drawing* is a correspondence sending vertices to points of $\mathbb{R}^2$ and edges to continuous curves between the corresponding vertices such that no two edges cross. An **infinite planar map** is an infinite planar graph equipped with a set of cyclic permutations $\{\sigma_v : v \in V\}$ of the neighbors of each vertex v, such that there exists a drawing of the graph which respects these permutations, that is, the clockwise order of edges emanating from a vertex v coincides with σ_v.

Unlike the finite case, one cannot define faces as the connected components of the plane with the edges removed since the drawing may have a complicated set of accumulation points. This is the reason that we have defined faces in Sect. 3.1 combinatorially, that is, based solely on the edge set and the cyclic permutation structure. This definition makes sense in both the finite and infinite case. In the latter case we may have infinite faces.

A (finite or infinite) planar map is a **triangulation** if each of its faces has exactly 3 edges. Given a drawing of a triangulation, the Jordan curve theorem implies that the edges of each face bound a connected component of the plane minus the edges. We will often refer to the faces as these connected components. A triangulation is called a **plane triangulation** if there exists a drawing of it such that every point of the plane is contained in either a face or an edge and any compact subset of the plane intersects at most finitely many edges and vertices. The term **disk triangulation** is also used in the literature and means the same with the unit disk taking the place of the plane in the previous definition. Of course these two definitions are equivalent since the plane and the open disk are homeomorphic. For example, take the product of the complete graph K_3 on 3 vertices with an infinite ray $\mathbb{N}$ and add a diagonal edge

© The Author(s) 2020
A. Nachmias, *Planar Maps, Random Walks and Circle Packing*, Lecture Notes in Mathematics 2243, https://doi.org/10.1007/978-3-030-27968-4_4

in each face that has 4 edges; this is a plane triangulation. However, the product of K_3 with a bi-infinite ray $\mathbb{Z}$ together with the same diagonals is a triangulation but not a plane triangulation, since it cannot be drawn in the plane without an accumulation point.

It turns out that there is a combinatorial criterion for a triangulation to be a plane/disk triangulation. We say that an infinite graph is **one-ended** if the removal of any finite set of its vertices leaves exactly one infinite connected component.

Lemma 4.1 *An infinite triangulation is a plane triangulation if and only if it is one-ended.*

Proof Suppose $G = (V, E)$ is a plane triangulation and consider a drawing of the graph with no accumulation points in the plane such that every point of the plane belongs to either an edge or a face. Let $A \subseteq V$ be a finite set of vertices and take $B \subset \mathbb{R}^2$ to be a ball around the origin which contains every vertex of A, every edge touching a vertex of A and every face incident to such an edge. Let $u \neq v$ be two vertices drawn outside of B and take a continuous curve γ between them in $\mathbb{R}^2 \setminus B$. By definition of B, this path only touches faces and edges that are not incident to the vertices of A and hence one can trace a discrete path from u to v in the graph that "follows" γ and avoids A. Since B intersects only finitely many edges and vertices, we learn that $G \setminus A$ has a unique infinite component.

Conversely, assume now that G is one-ended and consider a drawing of G in the plane. By the stereographic projection we project the drawing to the unit sphere $\mathbb{S}^2$ in $\mathbb{R}^3$. Denote by $\mathcal{I}$ the complement in $\mathbb{S}^2$ of the union of all faces and edges. Since G is an infinite triangulation this union is an open set, hence $\mathcal{I}$ is a closed set and its boundary $\partial \mathcal{I}$ is precisely the set of accumulation points of the drawing. Since $\mathcal{I}$ is closed, each connected component of $\mathcal{I}$ must be closed as well and hence contain at least one accumulation point. Since G is one-ended $\mathcal{I}$ cannot have more than one connected component, since otherwise we would be able to separate the two components by a finite set of edges and obtain two infinite connected components. Now choose a point $p \in \mathcal{I}$ and rotate the sphere so that p is the north pole. Project back the rotated sphere to the plane and consider the drawing in the plane. In this drawing the union of all faces and edges must be a simply connected set. By the Riemann mapping theorem this set is homeomorphic to the whole plane, and we deduce that the triangulation is a plane triangulation.   □

4.2   The Ring Lemma and Infinite Circle Packings

The circle packing theorem Theorem 3.5 is stated for finite planar maps. However, it is not hard to argue that any infinite map also has a circle packing. To this aim we will prove what is known as Rodin and Sullivan's *Ring Lemma* [70]; we will use it many times throughout this book. Given circles $C_0, C_1, \ldots, C_M$ with disjoint interiors, we say that $C_1, \ldots, C_M$ completely surround C_0 if they are all tangent to C_0 and C_i is tangent to C_{i+1} for $i = 1, \ldots, M$ (where C_{M+1} is set to be C_1).

Fig. 4.1 C_2 is small, but both
C_1 and C_3 are large

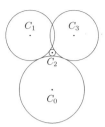

Lemma 4.2 (Ring Lemma, Rodin and Sullivan [70])   *For every integer $M > 0$ there exists $A > 0$ such that if C_0 is a circle completely surrounded by M circles $C_1, \ldots, C_M$, and r_i is the radius of C_i for every $i = 0, 1, \ldots, M$, then $r_0/r_i \leq A$ for every $i = 1, \ldots, M$.*

Proof We may scale the picture so that $r_0 = 1$. Assume that the radius of C_2 is small and consider the circles C_1 and C_3 to its left and right. It cannot be that both C_1 and C_3 have large radii compared to C_2 since in this case they will intersect; see Fig. 4.1. Hence, one of them has to be small as well. Assume without loss of generality that it is C_3. By similar reasoning, one of C_1 and C_4 has to be small. We continue this argument this way and get a path of circles of small radii; thus, for the circles $C_1, \ldots, C_M$ to completely surround C_0 we learn that M must be large.   □

For a circle packing P and a vertex v, denote by C_v the circle corresponding to v, by cent(v) the center of that circle, and by rad(v) its radius. We write $G(P)$ for the tangency graph of the packing P, that is, the graph in which each vertex is a circle of P and two such circles form an edge when they are tangent.

Claim 4.3 Let G be an infinite simple planar map. Then there exists a circle packing P such that $G(P)$ is isomorphic to G as planar maps.

Proof If G is not a triangulation, then it is always possible to add in each face new vertices and edges touching them so the resulting graph is a planar triangulation (in an infinite face we have to put infinitely many vertices). After circle packing this new graph, we can remove all the circles corresponding to the added vertices and remain with a circle packing of G. Thus, we may assume without loss of generality that G is a triangulation.

Fix a vertex x, and let G_n be the graph distance ball of radius n around x. Apply the circle packing theorem to G_n to obtain a packing P_n, and scale and translate it so that rad$(x) = 1$ and cent(x) is the origin.

Consider a neighbor y of x. By the Ring Lemma (Lemma 4.2), there exists a constant $A = A(x, y) > 0$ such that $A^{-1} \leq$ rad$(y) \leq A$. By compactness there exists a subsequence of packings P_{n_k} for which rad$_{n_k}(y)$ and cent$_{n_k}(y)$ both converge. By taking further subsequences for the rest of x's neighbors, and then for the rest of the graph's vertices, it follows by a diagonalization argument that there exists a subsequence such that the radii and centers of all vertices converge. The limiting packing P_∞ satisfies that $G(P_\infty)$ is isomorphic to G.   □

4.3   Statement of the He–Schramm Theorem

Given a circle packing P of a triangulation G, we define the **carrier** of P, denoted
Carrier(P), to be the union of the closed discs bounded by the circles of P together
with the spaces bounded between any three circles that form a face (i.e., the
interstices). When P is a circle packing of an infinite one-ended triangulation, the
argument in Lemma 4.1 shows that Carrier(P) is simply connected.

We say that G **is circle packed in** $\mathbb{R}^2$ when Carrier(P) $= \mathbb{R}^2$. Denote by $\mathbb{U}$ the
disk $\{z \in \mathbb{R}^2 : |z| < 1\}$; we say that G **is circle packed in** $\mathbb{U}$ when Carrier(P) $= \mathbb{U}$.
See Fig. 4.2.

Let G be a plane triangulation. Then G can be drawn in the plane $\mathbb{R}^2$ or
alternatively in the disk $\mathbb{U}$ (since they are homeomorphic), but can it be *circle
packed* both in $\mathbb{R}^2$ and in $\mathbb{U}$? A celebrated theorem of He and Schramm [40] states
that this cannot be done: each plane triangulation can be circle packed in either
the plane or the disk, but not both. In fact, the combinatorial property of G that
determines on which side of the dichotomy we are is the recurrence or transience of
the simple random walk on G (assuming also that G has bounded degrees, that is,
$\sup_{x \in V(G)} \deg(x) < \infty$). This is the content of the He–Schramm theorem, which
we are now ready to state.

Theorem 4.4 (He and Schramm [40]) *Let G be an infinite simple plane triangulation with bounded degrees.*

1. *If G is recurrent, then there exists a circle packing P of G such that
   Carrier(P) $= \mathbb{R}^2$.*
2. *If G is transient, then there exists a circle packing P of G such that
   Carrier(P) $= \mathbb{U}$.*

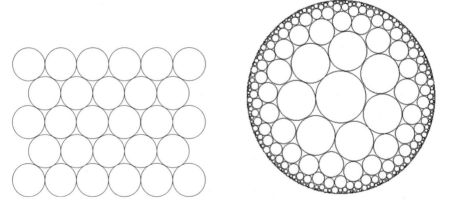

Fig. 4.2 Two circle packings with carriers $\mathbb{R}^2$ (left) and $\mathbb{U}$ (right)

3. *If P is a circle packing of G with* Carrier$(P) = \mathbb{R}^2$, *then G is recurrent.*
4. *If P is a circle packing of G with* Carrier$(P) = \mathbb{U}$, *then G is transient.*

Remark 4.5   Schramm [73] proved that a circle packing P of a triangulation $G(P)$ with Carrier$(P) = \mathbb{R}^2$ is uniquely determined up to dilations, rotations and translations. If Carrier$(P) = \mathbb{U}$ the same holds up to Möbius transformations of $\mathbb{U}$ onto itself (see also [37]). Hence the packings guaranteed to exist in Theorem 4.4 (1) and Theorem 4.4 (2) are unique in this sense.

Corollary 4.6   *Any bounded degree plane triangulation can be circle-packed in $\mathbb{R}^2$ or $\mathbb{U}$, but not both.*

Remark 4.7   In fact, it is proved in [40] that the corollary above holds without the assumption of bounded degree. Furthermore, in [40] Theorem 4.4 (1) and Theorem 4.4 (4) are proved without the bounded degrees assumption, but the other two statements require this assumption.

The following example demonstrates why the bounded degree condition is necessary for Theorem 4.4 (2) and Theorem 4.4 (3).

Example 4.8   Let P be a triangular lattice circle packing (as in Fig. 4.3), and let $C_0, C_1, C_2, \ldots$ be an infinite horizontal path of circles in P going (say) to the right. In the upper face shared by C_n and C_{n+1}, draw 2^n circles which form a vertical path and each of them tangent both to C_n and C_{n+1}; the last circle of these is also tangent to the upper neighbor of C_n and C_{n+1}. See Fig. 4.3.

The resulting graph is a plane triangulation and the carrier of the packing is $\mathbb{R}^2$. However, it is an easy exercise to verify that the tangency graph of this circle packing is transient.

In the rest of this chapter we prove Theorem 4.4. We begin by proving parts 3 and 4, in which a circle packing is given and we use its geometry to estimate

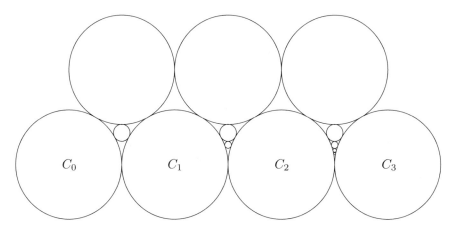

Fig. 4.3   Unbounded degree transient triangulation circle packed in $\mathbb{R}^2$

certain effective resistances. Afterwards we prove parts 1 and 2, in which we use
the electrical estimates to deduce facts about the geometry of the circle packing.

4.4   Proof of the He–Schramm Theorem

Proof of Theorem 4.4 (3)

Denote the circle packing $P = \{C_v\}_{v \in V}$ where V is the vertex set of G and C_v
denotes the circle corresponding to the vertex v. Write Δ for the maximum degree
of G and fix a vertex v_0. By scaling and translating we may assume that C_{v_0} is a
radius 1 circle around the origin. For a real number $R > 0$, let $V_R = V_{B(\mathbf{0},R)}$ denote
set of vertices v for which cent(v) is in the Euclidean ball of radius R around the
origin.

Lemma 4.9 *There exist $C = C(\Delta) > 1$ and $c = c(\Delta) > 0$ such that for every*
$R \geq 1$ *we have*

(i) *There are no edges between V_R and $V \setminus V_{CR}$, and*
(ii) $\mathcal{R}_{\mathrm{eff}}(V_R \leftrightarrow V \setminus V_{CR}) \geq c.$

Proof We begin with part (i). For every $v \in V_R$ it holds that rad$(v) \leq R$ since C_{v_0}
is centered at the origin. By the Ring Lemma (Lemma 4.2), there exists $A = A(\Delta)$
such that rad$(u) \leq AR$ for every $u \sim v$, and therefore $|\,\mathrm{cent}(u)| \leq (A + 2)R$.
Hence (i) holds with $C = A + 2$.

   To prove part (ii) we define

$$h(v) = \begin{cases} 0 & v \in V_R, \\ 1 & v \in V \setminus V_{CR}, \\ \frac{|\,\mathrm{cent}(v)| - R}{(C-1)R} & \text{otherwise.} \end{cases}$$

Recall from Lemma 2.32 that $\mathcal{R}_{\mathrm{eff}}(V_R \leftrightarrow V \setminus V_{CR}) \geq \mathcal{E}(h)^{-1}$. By the triangle
inequality, for an edge $\{x, y\}$ with both endpoints in $V_{CR} \setminus V_R$ we have

$$|h(x) - h(y)| \leq \frac{|\,\mathrm{cent}(x) - \mathrm{cent}(y)|}{(C-1)R} = \frac{\mathrm{rad}(x) + \mathrm{rad}(y)}{(C-1)R},$$

and it is straightforward to check that the same bound holds also when one of the
edge's endpoints is in V_R or $V \setminus V_{CR}$. Thus, using the Ring Lemma's (Lemma 4.2)
constant $A = A(\Delta)$ from part (i),

$$\mathcal{E}(h) \leq \sum_{x \in V_{CR} \setminus V_R} \sum_{y:y \sim x} \frac{((A+1)\,\mathrm{rad}(x))^2}{(C-1)^2 R^2} \leq \frac{\Delta(A+1)^2}{\pi(C-1)^2 R^2} \cdot \sum_{x \in V_{CR} \setminus V_R} \mathrm{area}(C_x),$$

where area(C_x) is the area that C_x encloses (that is, $\pi \operatorname{rad}(x)^2$). We have that $\sum_x \operatorname{area}(C_x) \leq \operatorname{area}(B(\mathbf{0}, 2CR)) = 4\pi C^2 R^2$, hence if $C = A + 2$, then

$$\mathcal{E}(h) \leq 4\Delta C^2,$$

and the result follows for $c = (4\Delta C^2)^{-1}$.   □

Proof of Theorem 4.4 (3) Consider the unit current flow I from v_0 to ∞ and fix any $R \geq 1$. Restricting this flow to the edges which have at least one endpoint in the annulus $V_{CR} \setminus V_R$ gives a unit flow from V_R to $V \setminus V_{CR}$, by part (i) of Lemma 4.9. Hence, by part (ii) of that lemma and by Thomson's principle (Theorem 2.28), the energy contributed to $\mathcal{E}(I)$ from these edges is at least c. In the same manner, the edges which have at least one endpoint in the annulus $V_{C^{2k+1}R} \setminus V_{C^{2k}R}$ contribute at least c to $\mathcal{E}(I)$. Part (i) of Lemma 4.9 implies that all these edge sets are disjoint, hence $\mathcal{E}(I) = \infty$ and we learn that G is recurrent (Corollary 2.39).   □

Proof of Theorem 4.4 (4)

We will use the given circle packing of G to create a random path to infinity with finite energy. This gives transience by Claim 2.46. This proof strategy is similar to that of Theorem 2.47.

Proof of Theorem 4.4 (4) Let v_0 be a fixed vertex of the graph, and apply a Möbius transformation to make the circle of P corresponding to v_0 be centered at the origin $\mathbf{0}$. We now use Claim 2.46 to construct a flow θ from v_0 to ∞ by choosing a uniform random point $\mathbf{p}$ on $\partial\mathbb{U}$, taking the straight line from $\mathbf{0}$ to $\mathbf{p}$ and considering the set of all circles in the packing P that intersect this line in the order that they are visited; this set forms an infinite simple path in the graph which starts at v_0.

To bound the energy of the flow, we claim that there exists some constant C (which may depend on the graph G and the packing P) such that the probability that the random path uses the vertex v is bounded above by $C \operatorname{rad}(v)$. Indeed, since there are only finitely many vertices with centers at distance at most $1/2$ from $\mathbf{0}$, we may assume that the center of v is of distance at least $1/2$ from $\mathbf{0}$. In this case, in order for v to be included in the random path the circle of v must intersect the line between $\mathbf{0}$ and $\mathbf{p}$. By the Ring Lemma (Lemma 4.2) the neighbors of v have circles of radii comparable to $\operatorname{rad}(v)$ and so the probability of the line touching them is at most $C \operatorname{rad}(v)$. Since the vertex degree is bounded by Δ and $\sum_{v \in V} \pi \operatorname{rad}(v)^2$ is at most the area of $\mathbb{U}$, we find that

$$\mathcal{E}(\theta) \leq C\Delta \sum_{v \in V} \operatorname{rad}(v)^2 \leq C\Delta.$$

Hence G is transient by Corollary 2.39   □

Proof of Theorem 4.4 (1)

We apply Claim 4.3 to obtain a circle packing P of G and prove that Carrier(P) = $\mathbb{R}^2$. Fix some vertex v and rescale and translate so that $P(v)$ is the unit circle $\partial \mathbb{U}$. Assume by contradiction that Carrier(P) $\neq \mathbb{R}^2$ and let $p \in \mathbb{R}^2 \setminus$ Carrier(P) be a point not in the carrier. Rotate the packing so that $p = R$ for some real number $R > 1$. Let $U \in [-1, 1]$ and consider the circle $C_U = \{z : |z - p| = R - U\}$. We traverse C_U from the point U counterclockwise and consider all the circles of P which intersect C_U. These circles form a simple path in the graph G starting from v. Since Carrier(P) is simply connected by Lemma 4.1 and $p \notin$ Carrier(P) it cannot be that $C_U \subset$ Carrier(P). Thus, as we traverse C_U counterclockwise we must hit the boundary of Carrier(P). We conclude that the path in G we obtained in this manner is an infinite simple path starting at v.

We now let U be a uniform random variable in $[-1, 1]$ and let μ denote the probability measure on random infinite paths starting at v we obtained as described above. Let θ be the flow induced by μ as in Claim 2.46. We wish to bound the energy $\mathcal{E}(\theta)$. Consider a vertex $w \in G$ and its corresponding circle C_w and let B be the Euclidean ball of radius $R + 1$ around p. If C_w does not intersect B, it cannot be included in the random path by our construction. If it does intersect this ball, then the probability that the random path intersects it is bounded above by its radius. Thus as in the proof of Theorem 4.4 (4),

$$\mathcal{E}(\theta) \leq C\Delta \sum_{w:C_w \cap B \neq \emptyset} \text{rad}(w)^2,$$

where Δ is the maximal degree of G and we have used the Ring Lemma (Lemma 4.2). We learn that $\mathcal{E}(\theta)$ is bounded above by a constant multiple of the area of all circles of P that intersect B. Since $p \notin$ Carrier(P), by the Ring Lemma (Lemma 4.2), any circle of P that intersects B cannot have radius more than AR for some large $A \geq R$ (since otherwise, all the circles surrounding this vertex will have radius more than $R + 1$, contradicting the fact that $p \notin$ Carrier(P)). We learn that all the circles counted in the sum above are contained in the Euclidean ball of radius $(A + 1)R + 1$ around p. Since these circles has disjoint interiors, the sum of their area is bounded above by the area of the Euclidean ball above. We conclude that $\mathcal{E}(\theta) < \infty$, hence G is transient by Corollary 2.39 and we have reached a contradiction.   □

Proof of Theorem 4.4 (2)

We will use the following simple corollary of the circle packing theorem, Theorem 3.5. A **finite triangulation with boundary** is a finite connected simple planar map in which all faces are triangles except for a distinguished outer face whose boundary is a simple cycle.

Claim 4.10 Let G be a finite triangulation with boundary. Then, there is a circle packing P of G such that all circles of the outer face are internally tangent to $\partial \mathbb{U}$ and all other circles of P are contained in $\mathbb{U}$.

Proof Denote by $v_1, \ldots, v_m$ the vertices of the outer face ordered according to the cycle they form. Add a new vertex v^* to the graph and connect it to $v_1, \ldots, v_m$ according to their order. We obtain a finite triangulation G^*. Apply Theorem 3.5 to obtain a circle packing $P = \{C_v\}_{v \in V(G^*)}$. By translating and dilating we may assume that C_{v^*} is centered at the origin and has radius 1. Apply the map $z \mapsto \frac{1}{z}$ on this packing. Since this map preserves circles, the image of the circles $\{C_v\}_{v \in V(G^*) \setminus \{v^*\}}$ under this map is precisely the desired circle packing.   □

Furthermore, we will require an auxiliary general estimate. Given a circle packing P and a set of vertices A, we write $\mathrm{diam}_P(A)$ for the Euclidean diameter of the union of all circles in P corresponding to the vertices of A.

Lemma 4.11 *Let P be a circle packing contained in $\mathbb{U}$ of a finite triangulation with boundary with maximum degree Δ, such that the circle of a chosen non-boundary vertex v_0 is centered at the origin and has radius r_0. Assume that $r_0 \geq r_{\min}$ for some constant $r_{\min} > 0$. Then there exists a constant $c = c(r_{\min}, \Delta) > 0$ such that for any connected set A of vertices,*

$$\mathcal{R}_{\mathrm{eff}}(v_0 \leftrightarrow A) \geq c \log \frac{1}{\mathrm{diam}_P(A)}. \tag{4.1}$$

If in addition all circles of the outer face are tangent to $\partial \mathbb{U}$ and A contains a vertex of the outer face, then

$$\mathcal{R}_{\mathrm{eff}}(v_0 \leftrightarrow A) \leq c^{-1} \log \frac{1}{\mathrm{diam}_P(A) \wedge \frac{1}{2}}. \tag{4.2}$$

Proof Write $\varepsilon = \mathrm{diam}_P(A)$ and let $z(A)$ denote the union of all circles corresponding to the vertices of A. We begin with the proof of (4.1), which goes along similar lines to the proof of Lemma 4.9. Let $z_0 \in \mathbb{R}^2$ be such that $z(A) \subset \{|z - z_0| \leq \varepsilon\}$. For any $r > 0$ denote by V_r the set of vertices whose corresponding circles have centers inside $\{|z - z_0| \leq r\}$, so that $A \subset V_\varepsilon$. Repeating the proof of Lemma 4.9 shows that there exists a constant $C = C(\Delta) > 0$ such that

(i) There are no edges between V_r and $V \setminus V_{Cr}$, and
(ii) $\mathcal{R}_{\mathrm{eff}}(V_r \leftrightarrow V \setminus V_{Cr}) \geq C^{-1}$, as long as V_r and $V \setminus V_{Cr}$ are non-empty.

Regarding this proof, we note that it is possible that the set $\{|z - z_0| \leq r\}$ is not contained in $\mathbb{U}$ (unlike the proof of Lemma 4.9 when the carrier is all of $\mathbb{R}^2$), however, this only works in our favor. The proof of (4.1) now proceeds similarly to the proof of Theorem 4.4 (3). When ε is small enough (depending only on $r_{\min}$

and Δ), by the Ring Lemma (Lemma 4.2), the Euclidean distance between the circle corresponding to v_0 and A is at least some constant (which again depends only on r_{min} and Δ) so that $v_0 \notin V_{CK\varepsilon}$ for some $K = \Omega(\log(1/\varepsilon))$. For each $k = 0, 2, 4, \ldots, K$ the sets of edges which have at least one endpoint in the annulus $V_{C^{k+1}\varepsilon} \setminus V_{C^k\varepsilon}$ are disjoint by (i). By (ii), each of these sets of edges contribute at least C^{-1} to the energy of the unit current flow from A to v_0, concluding the proof of (4.1) using Thomson's principle (Theorem 2.28).

For the proof of (4.2) we construct a unit flow from v_0 to A that has energy $O(\log(1/\varepsilon))$. The construction is in the same spirit as the proof of Theorem 4.4 (4), but there are some technical difficulties to overcome. Since A contains a vertex that is tangent to $\partial \mathbb{U}$, we choose $z_0 \in \partial \mathbb{U}$ that belongs to a circle of A. By rotating the packing we may assume that $z_0 = e^{i\varepsilon/4}$.

We now treat two cases separately. In the first case we assume that there exists z_1 in $z(A)$ such that $\arg(z_1) \in [0, \varepsilon/2]$ and $|z_1| \leq 1 - \varepsilon/2$ such that the path in $z(A)$ from z_0 to z_1 remains in the sector $\arg(z) \in [0, \varepsilon/2]$. Consider the points

$$x_0 = -r_0 \qquad x_1 = r_0 \qquad y_1 = 1 - \varepsilon/3 \qquad y_0 = 1 \,,$$

and note that x_0, x_1 are the leftmost and rightmost points on the circle of v_0. Let C_0 and C_1 be the upper half plane semi-circles in which x_0, y_0 and x_1, y_1 are antipodal points, respectively. The choice of y_0, y_1 is made so that the path between z_0 to z_1 in $z(A)$ must cross the region bounded by C_0, C_1 and the intervals $[x_0, x_1], [y_1, y_0]$, by our assumption on z_1 as long as ε is small enough. See Fig. 4.4, left.

For each $t \in [0, 1]$ write C_t for the upper half plane semi-circle in which $ty_1 + (1-t)y_0$ and $tx_1 + (1-t)x_0$ are antipodal points, so that C_t continuously interpolates between C_0 and C_1. See Fig. 4.4, left. Choose $t \in [0, 1]$ uniformly at random and consider the random path γ which traces C_t from left to right. This random path

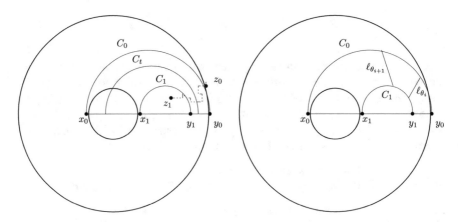

Fig. 4.4 Left: for any $t \in [0, 1]$ the semi-circle C_t must intersect the path in A between z_0 and z_1. Right: the quadrilateral Q_i is bounded between $\ell_{\theta_i}, \ell_{\theta_{i+1}}, C_0$ and C_1

starts at the circle of v_0 and must hit the path between z_0 and z_1 by our previous discussion. Hence, the circles of P that intersect γ must contain a path in the graph from v_0 to A. By Claim 2.46 we obtain a flow I from v_0 to A whose energy $\mathcal{E}(I)$ we now bound.

For an angle $\theta \in [0, \pi]$ we denote by $w_\theta(t)$ the point at angle θ, seen from the center of C_t, on the semi-circle C_t. It is an exercise to see that the set of points $\{w_\theta(t) : t \in [0, 1]\}$ form a straight line interval ℓ_θ. Furthermore, when t is chosen uniform in $[0, 1]$, the intersection of C_t and ℓ_θ is a uniformly chosen point on ℓ_θ. Set $\theta_0 = 0$ and $\theta_i = 2^{i-1}\varepsilon$ for $i = 1, \ldots, K-1$ where $K = O(\log(1/\varepsilon))$ such that $\theta_{K-1} \in [\pi/4, \pi/2]$ and set $\theta_K = \pi$. We will obtain the bound $\mathcal{E}(I) = O(K)$ by bounding from above by a constant the contribution to $\mathcal{E}(I)$ coming from edges which intersect the quadrilateral Q_i of $\mathbb{R}^2$ bounded by $\ell_{\theta_i}, \ell_{\theta_{i+1}}, C_0, C_1$; see Fig. 4.4, right. The random path γ restricted to Q_i can be sampled by choosing a uniform random point on ℓ_{θ_i}, setting $t \in [0, 1]$ to be the unique number such that C_t intersects ℓ_{θ_i} at the chosen point, and tracing the part of C_t from ℓ_{θ_i} to $\ell_{\theta_{i+1}}$. The lengths of the four curves bounding Q_i are all of order $2^i \varepsilon$ and so we deduce that if v corresponds to a circle of radius $O(2^i \varepsilon)$ which intersects Q_i, then the probability that it is visited by γ is $O(\mathrm{rad}(v)/2^i \varepsilon)$. Since the sum of $\mathrm{rad}(v)^2$ over such v's is at most the area of Q_i up to a multiplicative constant (note that some of these circles need not be contained in Q_i) it is at most $O(2^{2i}\varepsilon^2)$. Since the degrees are bounded we deduce that the contribution to the energy from edges touching such v's is $O(1)$. Lastly, if v corresponds to a larger circle, then we bound its probability of being visited by γ by 1 and note that there can only be $O(1)$ many such v's whose circles intersects Q_i. Thus the contribution from these is another $O(1)$. Since there are $O(\log(1/\varepsilon))$ such i's we learn that $\mathcal{E}(I) = O(\log(1/\varepsilon))$ finishing our proof in this case using Thomson's principle (Theorem 2.28).

In the second case, we assume that there exists $z_1 \in z(A)$ such that $\arg(z_1) \notin [0, \varepsilon/2]$ and $|z_1| \geq 1 - \varepsilon$. It is clear that since $\mathrm{diam}_P(A) = \varepsilon$ either the first or the second case must occur. Denote $z_0' = |z_1|e^{i\varepsilon/4}$ and let x_0, x_1 be antipodal points on the circle of v_0 such that the straight line between x_0 and x_1 is parallel to the straight line between z_0' and z_1. The vertices z_0', z_1, x_0, x_1 form a trapezoid, see Fig. 4.5. We choose a uniform random point $t \in [0, 1]$ and stretch a straight line from $tx_0 + (1-t)x_1$ to $tz_0 + (1-t)z_1'$. We then continue it by a straight line from $tz_0 + (1-t)z_1'$ to $w \in \partial\mathbb{U}$ where $\arg(w) = \arg(tz_0 + (1-t)z_1')$. Denote the resulting path by γ_t and note that it starts inside the circle of v_0 and must hit the path between z_0 and z_1 in $z(A)$. Thus, the set of all circles which intersect γ_t form a path in the graph that starts at v_0 and ends at A; this random choice of γ_t gives us as usual a unit flow from v_0 to A by Claim 2.46. By repeating the same argument as in the previous case (that is, splitting the trapezoid into $O(\log(1/\varepsilon))$ many trapezoids of constant aspect ratio), we see that the contribution to the energy of the flow induced by the random path γ_t of the edges in the trapezoid is $O(\log(1/\varepsilon))$. Furthermore, the same argument gives that the edges in the quadrilateral formed by the vertices z_0, z_0', z_1 and $e^{i\,\arg(z_1)}$ also contribute at most a constant to the energy, concluding our proof for the second case by Thomson's principle again (Theorem 2.28).   □

Fig. 4.5 The resistance
across the trapezoid on
vertices x_0, x_1, z'_0, z_1 is
$O(\log(1/\varepsilon))$ when
$|z'_0 - z_1| = \Theta(\varepsilon)$

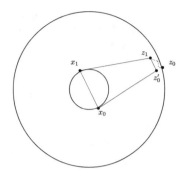

Proof of Theorem 4.4 (2) Denote by $d_G(u, v)$ the graph distance between the vertices u, v of G. Fix some $v_0 \in V$ and let

$$B_j = \{v : d_G(v_0, v) \leq j\},$$

$$V_j = B_j \cup \{\text{finite components of } V \setminus B_j\},$$

$$E_j = \{\text{edges induced by } V_j\}.$$

The graph $G_j = (V_j, E_j)$ with the map structure inherited from G is a finite triangulation with boundary. Indeed, it is straightforward to check that it is 2-connected (i.e., the removal of a single vertex does not disconnect the graph) which implies that the outer face forms a simple cycle, see [20, Proposition 4.2.5]. Furthermore, since G is one-ended and we have added all the finite components in $V \setminus B_j$ there cannot be a face with more than 3 edges except for the outer face which we denote by ∂G_j.

Thus G_j is an increasing sequence of finite triangulations with boundary such that $\cup_j G_j = G$. We apply Claim 4.10 to pack G_j inside the unit disk $\mathbb{U}$ such that the circles of ∂G_j are tangent to $\partial \mathbb{U}$. By applying a Möbius transformation from $\mathbb{U}$ onto $\mathbb{U}$, we may assume that the circle corresponding to v_0 is centered at the origin. We denote this packing by P_j and let r_0^j be the radius of v_0 in P_j.

Since G is transient it follows that there exists some $c = c(\Delta) > 0$ such that $r_0^j \geq c$ for all j by Corollary 2.39. Indeed, if $r_0^j \leq \varepsilon$, we learn by Lemma 4.9 and the proof of Theorem 4.4 (3) that $\mathcal{R}_{\text{eff}}(v_0 \leftrightarrow \infty) \geq c' \log(\varepsilon^{-1})$ for some $c' = c'(\Delta) > 0$.

As we did in Claim 4.3, we now take a subsequence in which the centers and radii of all vertices converge. Denote the resulting limiting packing by P_∞. This packing has all circles inside $\mathbb{U}$ and we therefore deduce that Carrier$(P_\infty) \subseteq \mathbb{U}$. It is a priori possible that Carrier(P_∞) is some strict subset of $\mathbb{U}$, i.e., that all the circles stabilize inside some strict subset of $\mathbb{U}$. We now argue that this is not possible.

Let Z be the set of accumulation points of Carrier(P_∞); it suffices to show that $Z \subset \partial \mathbb{U}$ since any simply connected domain $\Omega \subset \mathbb{U}$ for which $\partial \Omega \subset \partial \mathbb{U}$ must equal $\mathbb{U}$. Since Z is a compact set, let $z \in Z$ minimize $|z|$ among all $z \in Z$; it suffices to show that $z \in \partial \mathbb{U}$. Fix $\varepsilon > 0$ and put

$$U_\varepsilon(z) = \left\{ v \in G : |\operatorname{cent}_{P_\infty}(v) - z| \leq \varepsilon \right\}.$$

The graph spanned on the vertices $U_\varepsilon(z)$ may be disconnected, yet by our choice of z it is clear that $U_\varepsilon(z)$ contains an infinite connected component. Indeed, one can draw a straight line from the origin to z without intersecting Z and consider the set of all circles intersecting this line; from some point onwards the vertices corresponding to these circles will reside in $U_\varepsilon(z)$.

Therefore, let $W_\varepsilon(z)$ be an infinite connected component of the graph spanned on $U_\varepsilon(z)$. Let $J = J(z, \varepsilon)$ be the first integer such that $V_J \cap W_\varepsilon(z) \neq \emptyset$. Since the G_j's are increasing finite sets and $W_\varepsilon(z)$ is an infinite connected set, we have that $\partial G_j \cap W_\varepsilon(z) \neq \emptyset$ for all $j \geq J$. Consider now any connected component A_j of the graph spanned on the vertices $V_j \cap W_\varepsilon(z)$.

Denote by P_∞^j the finite circle packing obtained from P_∞ by taking only the circles of V_j. It has the same adjacency graph as P_j but it is a different packing. Since $A_j \subset W_\varepsilon(z)$, it follows that $\operatorname{diam}_{P_\infty^j}(A_j) \leq 4\varepsilon$. By Lemma 4.11, Eq. (4.1), applied to the set A_j in the packing P_∞^j, we deduce that $\mathcal{R}_{\text{eff}}(v_0 \leftrightarrow A_j; G_j) \geq c \log(1/\varepsilon)$. Since A_j is a connected component of $V_j \cap W_\varepsilon(z)$ and since $W_\varepsilon(z)$ is an infinite connected set of vertices in G, it follows that A_j must contain a vertex of ∂V_j. Thus, we may apply Lemma 4.11, Eq. 4.2, to the set A_j, this time in the packing P_j, to get that there exists some $c > 0$ such that

$$\operatorname{diam}_{P_j}(A_j) \leq \varepsilon^c. \tag{4.3}$$

Choose some $v_J \in \partial G_J \cap W_\varepsilon(z)$ so that $|\operatorname{cent}_{P_\infty}(v_J) - z| \leq \varepsilon$. For each $j \geq J$ choose $v_j \in \partial G_j \cap W_\varepsilon(z)$ so that v_j and v_J are in the same connected component A_j of the graph spanned on $V_j \cap W_\varepsilon(z)$. Since the circle of v_j in P_j touches $\partial \mathbb{U}$ we learn by (4.3) that the distance of the circle of v_J in P_j from $\partial \mathbb{U}$ is at most ε^c for all $j \geq J$. Since the circle corresponding to v_J in P_∞ is the limit of its circles in P_j we deduce that the distance of $\operatorname{cent}_{P_\infty}(v_J)$ from $\partial \mathbb{U}$ is at most ε^c. Hence the distance of z from $\partial \mathbb{U}$ is at most $\varepsilon + \varepsilon^c$. Since ε was arbitrary we obtain that $z \in \partial \mathbb{U}$, as required. □

4.5 Exercises

1. Let G be a triangulation of the plane with maximal degree at most 6. Prove that the simple random walk on G is recurrent.
2. Let G be a plane triangulation that can be circle packed in the unit disc $\{z : |z| < 1\}$. Show that the simple random walk on G is transient. (Note that G may have *unbounded* degrees)

3.(*) Let P be a circle packing of a finite simple planar map with degree bounded by D such that all of its faces are triangles except for the outerface. Assume that the carrier of P is contained in $[-11, 11]^2$, contains $[-10, 10]^2$ and that all circles have radius at most 1. Let h be the harmonic function taking the value 1 on all vertices with centers left of the line $\{-10\} \times \mathbb{R}$, taking the value 0 on all vertices with centers right of the line $\{10\} \times \mathbb{R}$, and is harmonic anywhere else. Assume x and y are two vertices such that their centers are contained in $[-1, 1]^2$ and that the Euclidean distance between these centers is at most $\epsilon > 0$. Show that

$$|h(x) - h(y)| \leq \frac{C}{\log(1/\epsilon)},$$

for some constant $C = C(D) > 0$ independent of ϵ. [Hint: assume $h(x) < h(y)$ and consider the sets $A = \{v : h(v) \leq h(x)\}$ and $B = \{v : h(v) \geq h(y)\}$].

Chapter 5
Planar Local Graph Limits

5.1 Local Convergence of Graphs and Maps

In order to study large random graphs it is mathematically natural and appealing to introduce an infinite limiting object and study its properties. In their seminal paper, Benjamini and Schramm [11] introduced the notion of locally convergent graph sequences, which we now describe.

We will consider random variables taking values in the space $\mathcal{G}_\bullet$ of locally finite connected rooted graphs viewed up to root preserving graph isomorphisms. That is, $\mathcal{G}_\bullet$ is the space of pairs (G, ρ) where G is a locally finite graph (which may be finite or infinite) and $\rho \in V(G)$ is a vertex of G and two elements $(G_1, \rho_1), (G_2, \rho_2)$ are considered equivalent if there is a graph isomorphism between them (that is, a bijection $\varphi : V(G_1) \to V(G_2)$ such that $\varphi(\rho_1) = \varphi(\rho_2)$ and $\{v_1, v_2\} \in E(G_1)$ if and only if $\{\varphi(v_1), \varphi(v_2)\} \in E(G_2)$). We remark that throughout this book our graphs will almost entirely be connected. In the rare case when G is not connected, we impose the convention that (G, ρ) refers to $(G[\rho], \rho)$ where $G[\rho]$ is the connected component of ρ in G. This way $(G, \rho) \in \mathcal{G}_\bullet$ even when G is disconnected (this will only be relevant in Chap. 6, and in particular in Lemma 6.11 and its usage).

In a similar fashion we define $\mathcal{M}_\bullet$ to be the set of equivalence classes of rooted maps; in this case we require the graph isomorphism to preserve additionally the cyclic permutations of the neighbors of each vertex, that is, it is a **map isomorphism**. Let us describe the topology on $\mathcal{G}_\bullet$ and $\mathcal{M}_\bullet$. For convenience we discuss $\mathcal{G}_\bullet$ but every statement in the following holds for $\mathcal{M}_\bullet$ as well.

Given an element (G, ρ) of $\mathcal{G}_\bullet$, the finite graph $B_G(\rho, R)$ is the subgraph of (G, ρ) rooted at ρ spanned by the vertices of distance at most R from ρ. We provide $\mathcal{G}_\bullet$ with a metric d_{loc}

$$d_{\mathrm{loc}}((G_1, \rho_1), (G_2, \rho_2)) = 2^{-R},$$

© The Author(s) 2020
A. Nachmias, *Planar Maps, Random Walks and Circle Packing*, Lecture Notes in Mathematics 2243, https://doi.org/10.1007/978-3-030-27968-4_5

where R is the largest integer for which $B_{G_1}(\rho_1, R)$ and $B_{G_2}(\rho_2, R)$ are isomorphic as graphs. This is a separable topological space (the finite graphs form a countable base for the topology) and is easily seen to be complete, thus it is a Polish space. The distances are bounded by 1 but the space is not compact. Indeed, the sequence G_n of stars with n leaves emanating from the root ρ has no converging subsequence.

Since $\mathcal{G}_\bullet$ is a Polish space, we can discuss convergence in distribution of a sequence of random variables $\{X_n\}_{n=1}^\infty$ taking values in $\mathcal{G}_\bullet$. We say that X_n **converges in distribution** to a random variable X, and denote it by $X_n \xrightarrow{d} X$, if for every bounded continuous function $f : \mathcal{G}_\bullet \to \mathbb{R}$ we have that $\mathbb{E}(f(X_n)) \to \mathbb{E}(f(X))$. We will be focused here on the particular situation in which X_n is a *finite* rooted random graph (G_n, ρ_n) such that given G_n, the root ρ_n is uniformly distributed among the vertices of G_n. It is a very common setting and justifies the following definition.

Definition 5.1 Let $\{G_n\}$ be a sequence of (possibly random) finite graphs. We say that G_n **converges locally** to a (possibly infinite) random rooted graph $(U, \rho) \in \mathcal{G}_\bullet$, and denote it by $G_n \xrightarrow{\text{loc}} (U, \rho)$, if for every integer $r \geq 1$,

$$B_{G_n}(\rho_n, r) \xrightarrow{d} B_U(\rho, r),$$

where ρ_n is a uniformly chosen vertex from G_n.

It is straightforward to see that this definition is equivalent to saying that the random variables (G_n, ρ_n) converge in distribution to (U, ρ). Note that this definition is consistent whether G_n is a deterministic finite graph or is a random variable drawn from some probability measure. In both cases $B_{G_n}(\rho_n, r)$ is a random variable taking values in $\mathcal{G}_\bullet$ and we clarify that ρ_n is drawn uniformly *conditioned* on G_n.

Examples

- The sequence $\{G_n\}$ of paths of length n converges locally to the graph $(\mathbb{Z}, 0)$ (note that the root vertex can be chosen to be any vertex of $\mathbb{Z}$ since $(\mathbb{Z}, i)$ and $(\mathbb{Z}, j)$ are equivalent for all $i, j \in \mathbb{Z}$).
- The sequence $\{G_n\}$ of the $n \times n$ square grid converges locally to the graph $(\mathbb{Z}^2, \mathbf{0})$ (again the root can be chosen to be any vertex of $\mathbb{Z}^2$).
- Let $\lambda > 0$ be fixed and let $\{G(n, \frac{\lambda}{n})\}$ be the sequence of random graphs obtained from the complete graph K_n by retaining each edge with probability $\frac{\lambda}{n}$ and erasing it otherwise, independently for all edges. This is known as the Erdös-Rényi random graph. One can verify that this sequences converges locally to a branching process with progeny distribution Poisson(λ). See exercise 1 of this chapter.

Fig. 5.1 A part of the
canopy tree

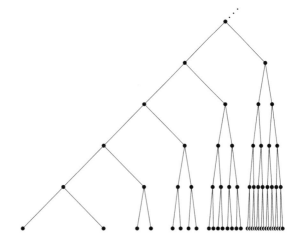

- Let G_n be the binary tree of height n. Perhaps surprisingly, its local limit is *not* the infinite binary tree. Instead, it is the following so-called **canopy tree** depicted in Fig. 5.1 and the root is at distance $k \geq 0$ from the leaves with probability 2^{-k-1}. Note that the distance of the root from the leaves determines the isomorphism class of the rooted graph. It is easy to see that the canopy tree is not isomorphic to the infinite binary tree, for example, it has leaves; furthermore, unlike the infinite binary tree it is recurrent.
- Consider G_n to be a path of length n, glued via one of its leaves into a $\sqrt{n} \times \sqrt{n}$ grid. The local limit of G_n is (U, ρ), where (U, ρ) is $(\mathbb{Z}, 0)$ with probability $1/2$, and $(\mathbb{Z}^2, \mathbf{0})$ otherwise.

Our goal in this chapter is to prove the following pioneering result.

Theorem 5.2 (Benjamini–Schramm [11]) *Let $M < \infty$ and let G_n be finite planar maps (possibly random) with degrees almost surely bounded by M such that $G_n \xrightarrow{\text{loc}} (U, \rho)$. Then (U, ρ) is almost surely recurrent.*

For instance, a local limit of planar maps cannot be the 3-regular infinite tree (however, the 3-regular infinite tree can be obtained as a local limit of uniformly random 3-regular graphs). The bounded degree assumption in Theorem 5.2 is necessary. Indeed, suppose we start with a binary tree of height n and replace each edge (u, v) that is at distance $k \geq 0$ from the leaves by 2^k parallel edges. By the same reasoning of the local convergence of binary trees to the canopy tree, the modified graph sequence converges locally to a modified canopy tree in which an edge at distance k from the leaves is replaced with 2^k parallel edges. Using the parallel law it is immediate to see that this graph is transient, and that the effective resistance from a leaf to ∞ is at most 2 (in fact it equals 2). See Fig. 5.2.

Fig. 5.2 A part of a
transient canopy tree.
Numbers on edges are
conductances of those edges
after applying the parallel law

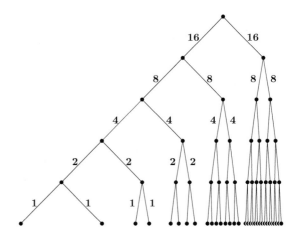

5.2   The Magic Lemma

Suppose $C \subseteq \mathbb{R}^2$ is finite. For each $w \in C$, define

$$\rho_w = \min\{|v - w| : v \in C \setminus \{w\}\}.$$

We call ρ_w the **isolation radius** of w. Given $\delta \in (0, 1)$, $s \geq 2$ and $w \in C$, we
say that w is (δ, s)-**supported** if in the disk of radius $\delta^{-1}\rho_w$ around w there are
at least s points of C outside any given disk of radius $\delta\rho_w$. In other words, w is
(δ, s)-**supported** if

$$\inf_{p \in \mathbb{R}^2} \left| C \cap B\left(w, \delta^{-1}\rho_w\right) \setminus B(p, \delta\rho_w) \right| \geq s.$$

The proof of Theorem 5.2 is based on the following lemma, which has been
dubbed "the Magic Lemma".

Lemma 5.3 ([11]) *There exists $A > 0$ such that for every $\delta \in (0, 1/2)$, every finite
$C \subseteq \mathbb{R}^2$ and every $s \geq 2$, the number of (δ, s)-supported points in C is at most*

$$\frac{A|C|\delta^{-2}\ln(\delta^{-1})}{s}.$$

Remark 5.4 We prove the lemma for $\mathbb{R}^2$, but it holds for $\mathbb{R}^d$ or any other doubling
metric space. In fact, a metric space for which the lemma holds must be doubling;
see [29].

Proof of Lemma 5.3

Let $k \geq 3$ be an integer (later we will take $k = k(\delta)$). Let G_0 be a tiling of $\mathbb{R}^2$ by 1×1 squares, rooted at some point p, and for every $n \in \mathbb{Z}$, let G_n be a tiling of $\mathbb{R}^2$ by $k^n \times k^n$ such that each square of G_n is tiled by $k \times k$ squares of G_{n-1}. We may choose p so that none of the points of C lies on the edge of a square.

We say that a square $S \in G_n$ is s-**supported** if for every smaller square $S' \in G_{n-1}$ we have that $|C \cap (S \setminus S')| \geq s$.

Claim 5.5 For any $s \geq 2$ the total number of s-supported squares, in $G = \bigcup_{n \in \mathbb{Z}} G_n$, is at most $2|C|/s$.

Proof Define a "flow" $f : G \times G \to \mathbb{R}$ as follows:

$$
f(S', S) = \begin{cases} \min(s/2, |S' \cap C|) & S' \subseteq S, \, S' \in G_n, \, S \in G_{n+1}, \\ -f(S, S') & S \subseteq S', \, S \in G_n, \, S' \in G_{n+1}, \\ 0 & \text{otherwise.} \end{cases}
$$

Let us make two initial observations. First we have that

$$
\sum_{S' \in G} f(S', S) \geq 0, \tag{5.1}
$$

by splitting into the two cases depending on whether there exists a square $S' \subseteq S$ such that $f(S', S) = s/2$ or not. Secondly, if S is a s-supported square

$$
\sum_{S' \in G} f(S', S) \geq \frac{s}{2}, \tag{5.2}
$$

by splitting into cases depending on whether the number of squares $S' \subseteq S$ such that $f(S', S) = s/2$ is at most one or at least two.

Let $a \in \mathbb{Z}$ be such that each square in G_a contains at most 1 point of C so there are no s-supported squares in $\bigcup_{n \leq a} G_n$. It easily follows from the definition of f that

$$
\sum_{S' \in G_a} \sum_{S \in G_{a+1}} f(S', S) = |C|, \tag{5.3}
$$

and that for every $b \in \mathbb{Z}$

$$
\sum_{S' \in G_b} \sum_{S \in G_{b+1}} f(S', S) \geq 0. \tag{5.4}
$$

Now, using (5.3) and (5.4),

$$\sum_{n=a+1}^{b}\sum_{S\in G_n}\sum_{S'\in G}f(S',S)=\sum_{n=a+1}^{b}\sum_{S\in G_n}\left(\sum_{S'\in G_{n-1}}f(S',S)+\sum_{S'\in G_{n+1}}f(S',S)\right)$$

$$=\sum_{S\in G_{a+1}}\sum_{S'\in G_a}f(S',S)+\sum_{S\in G_b}\sum_{S'\in G_{b+1}}f(S',S)\leq|C|.$$

Therefore, using (5.1) and (5.2), we deduce that there are at most $2|C|/s$ squares in $\bigcup_{b\geq n>a}G_n$ that are s-supported. Sending $b\to\infty$ finishes the proof. $\square$

The above claim is very close to the statement of Lemma 5.3 which we are pursuing. However, we need to move from squares to circles. We choose $k=\lceil20\delta^{-2}\rceil$ and let $\beta\sim\mathsf{Unif}([0,\ln k])$. Let G_0 be a tiling with side length e^β, based at the origin. Suppose we have defined G_n as a tiling of squares of side length $e^\beta k^n$; then G_{n+1} is a tiling of squares of side length $e^\beta k^{n+1}$ that is based uniformly at one of the k^2 possible points of G_n. Because the desired statement is invariant under translation and dilation of C, we may assume that C does not intersect the edges of G_n (for every n) and that $\rho_w\geq k$ for every $w\in C$. We call a point $w\in C$ a **city** in a square $S\in G$ if:

- the side length of S is in the interval $[4\delta^{-1}\rho_w,5\delta^{-1}\rho_w]$, and
- the distance from w to the center of S is at most $\delta^{-1}\rho_w$.

Claim 5.6 The probability that any given $w\in C$ is a city is $\Omega(\ln^{-1}(\delta^{-1}))$.

Proof For the first item to hold, β needs to satisfy that there exists $n\in\mathbb{Z}$ such that $e^\beta k^n\in[4\delta^{-1}\rho_w,5\delta^{-1}\rho_w]$, or $\beta+n\ln k\in\ln(\delta^{-1}\rho_w)+[\ln 4,\ln 5]$. Since $\beta\in\mathsf{Unif}([0,\ln k])$, the probability for that is $(\ln(5/4))/\ln k$, which is $\Omega(\ln^{-1}(\delta^{-1}))$ when $\delta\in(0,1/2)$.

As for the second item, it holds with positive probability (independent of δ) over the k^2 choices for basing G_n on top of G_{n-1}, given that β satisfies the requirement posed by the first item. $\square$

Claim 5.7 If w is a city in S and is (δ,s)-supported, then S is s-supported.

Proof If $S\in G_n$ is as above, then any little square $S'\in G_{n-1}$ has side length at most

$$\frac{\delta^2}{20}\cdot\frac{5\rho_w}{\delta}=\frac{\delta\rho_w}{4}.$$

Hence, it is contained in a disk of radius $\delta\rho_w$. Thus, for every $S'\in G_{n-1}$ with $S'\subseteq S$ there exists a point p such that

$$|C\cap(S\setminus S')|\geq\left|C\cap\left(B\left(w,\delta^{-1}\rho_w\right)\setminus B(p,\delta\rho_w)\right)\right|\geq s,$$

where we have used the fact that $B\left(w, \delta^{-1}\rho_w\right) \subset S$.   □

Now note that the expected number of pairs (w, S) such that S is s-supported, w is (δ, s)-supported, and w is a city, is at least $c \ln^{-1}(\delta^{-1}) N$, where N is the number of (δ, s)-supported points. Also, no more than $c\delta^{-2}$ points of C can be cities in a single square S. It follows from Claim 5.5 that

$$N \leq \frac{A|C|\delta^{-2} \ln(\delta^{-1})}{s},$$

concluding the proof of Lemma 5.3.   □

5.3 Recurrence of Bounded Degree Planar Graph Limits

Theorem 5.2 follows immediately from the following theorem which gives a quantitative estimate on the growth of the resistance in local limits of bounded degree planar maps. In particular, it states that the resistance grows logarithmically in the Euclidean distance of the corresponding circle packing.

Theorem 5.8 *Let (U, ρ) be a local limit of (possibly random) finite planar maps with maximum degree at most D. Then, almost surely, there exist a constant $c > 0$ and a sequence $\{B_k\}_{k \geq 1}$ of subsets of U such that for each k we have*

1. *$|B_k| \leq c^{-1}k$, and*
2. *$\mathcal{R}_{\mathrm{eff}}(\rho \leftrightarrow U \setminus B_k) \geq c \log k$.*

In particular, (U, ρ) is almost surely recurrent.

We write $B_{\mathrm{euc}}(p, r)$ for the Euclidean ball of radius r around a point $p \in \mathbb{R}^2$. As before, for a subset $O \subset \mathbb{R}^2$ and a given circle packing we write V_O for the set of vertices in which the centers of the corresponding circles are in O. In order to prove Theorem 5.8, we will need the following immediate corollary of the Magic Lemma (Lemma 5.3):

Corollary 5.9 *Let G be a finite simple planar triangulation, and P a circle packing of G. Let ρ be a uniform random vertex and P' a dilation and translation of P such that the circle of ρ is a unit circle centered at the origin $\mathbf{0}$. Then, there exists a universal constant $A > 0$ such that in the packing P', for every real $r \geq 2$ and integer $s \geq 2$*

$$\mathbb{P}\left(\forall p \in \mathbb{R}^2 \quad \left|V_{B_{\mathrm{euc}}(\mathbf{0},r) \setminus B_{\mathrm{euc}}(p,\frac{1}{r})}\right| \geq s\right) \leq \frac{Ar^2 \log r}{s}.$$

Proof Apply the Magic Lemma with $\delta = \frac{1}{r}$ and $s = s$, with the centers of circles of P' as the point set C. Note that there exists a constant $C > 0$ such that for all

$w \in V$ the isolation radius of w, ρ_w, satisfies $\mathrm{rad}(C_w) \le \rho_w \le C\,\mathrm{rad}(C_w)$ (without appealing to the Ring Lemma). $\qquad\qquad\qquad\qquad\qquad\qquad\qquad\qquad\qquad\qquad\qquad$ □

The following lemma provides the main estimate needed to prove Theorem 5.8. Once it has been shown, Theorem 5.8 will follow by a Borel-Cantelli argument.

Lemma 5.10 *Let G be a finite simple planar map with maximum degree at most D and let ρ be a uniform random vertex of G. Then, there exists a constant $C = C(D) < \infty$ such that for all $k \ge 1$,*

$$\mathbb{P}\left(\exists B \subseteq V, \ |B| \le Ck, \ \mathcal{R}_{\mathrm{eff}}(\rho \leftrightarrow V \setminus B) \ge C^{-1}\log k\right) \ge 1 - Ck^{-\frac{1}{3}}\log k,$$

where we interpret $\mathcal{R}_{\mathrm{eff}}(\rho \leftrightarrow V \setminus B) = \infty$ when $B = V$.

Proof We first assume that G is a triangulation and consider a circle packing of it where the circle of ρ is a unit circle centered at the origin $\mathbf{0}$. Applying Corollary 5.9 with $r = k^{\frac{1}{3}}$, $s = k$, we have that with probability at least $1 - Ak^{-\frac{1}{3}}\log(k)/3$, there exists $p \in \mathbb{R}^2$ with

$$\left| V_{B_{\mathrm{euc}}(\mathbf{0},r) \setminus B_{\mathrm{euc}}(p,\frac{1}{r})} \right| < k.$$

Now, if $|V_{B_{\mathrm{euc}}(p,\frac{1}{r})}| \le 1$, we set $B = V_{B_{\mathrm{euc}}(\mathbf{0},r)}$. We then have $|B| \le k$ and by applying $\Omega(\log k)$ times Lemma 4.9 together with the series law (Claim 2.24) we get that $\mathcal{R}_{\mathrm{eff}}(\rho \leftrightarrow V \setminus B) \ge c\log k$ for some $c = c(D) > 0$. Else, if $|V_{B_{\mathrm{euc}}(p,\frac{1}{r})}| \ge 2$ then we take $B = V_{B_{\mathrm{euc}}(\mathbf{0},r)} \setminus V_{B_{\mathrm{euc}}(p,\frac{1}{r})}$. By the Ring Lemma, there exists a $c' = c'(D) > 0$ such that $|p| \ge 1 + c'$. Since $|V_{B_{\mathrm{euc}}(p,\frac{1}{r})}| \ge 2$, we have a vertex in that set with radius at most r^{-1}. Therefore, $B_{\mathrm{euc}}(p, \frac{2}{r})$ contains at least one full circle C_v. Hence, by scaling and translating such that $C_v = \mathbb{U}$, we get (again, by Lemma 4.9) that

$$\mathcal{R}_{\mathrm{eff}}\left(V_{B_{\mathrm{euc}}(p,\frac{2}{r})} \leftrightarrow V \setminus V_{B_{\mathrm{euc}}(p,c'/2)}\right) \ge c_2\log k,$$

for some other constant $c_2 = c_2(D) > 0$. Since $\rho \in V \setminus V_{B_{\mathrm{euc}}(p,c'/2)}$ we obtain

$$\mathcal{R}_{\mathrm{eff}}\left(\rho \leftrightarrow V_{B_{\mathrm{euc}}(p,\frac{2}{r})}\right) \ge c_2\log k.$$

By Lemma 4.9 we also have that

$$\mathcal{R}_{\mathrm{eff}}\left(\rho \leftrightarrow V \setminus V_{B_{\mathrm{euc}}(\mathbf{0},r)}\right) \ge c_3\log k,$$

for some $c_3 = c_3(D) > 0$. By Claim 2.22 this means that

$$\mathbb{P}_\rho \left(\tau_{V \setminus V_{Beuc(0,r)}} < \tau_\rho^+ \right) \le \frac{1}{c_2 \log(k)} \quad \text{and} \quad \mathbb{P}_\rho \left(\tau_{V_{Beuc(p,\frac{2}{r})}} < \tau_\rho^+ \right) \le \frac{1}{c_3 \log(k)} .$$

By the union bound

$$\mathbb{P}_\rho \left(\tau_{V \setminus B} < \tau_\rho^+ \right) \le \frac{2}{\min(c_2, c_3) \log(k)} ,$$

hence by Claim 2.22 again

$$\mathcal{R}_{\text{eff}} \left(\rho \leftrightarrow V \setminus B \right) \ge \min(c_2, c_3) D^{-1} \log(k)/2 ,$$

concluding the proof when G is a triangulation.

If G is not a triangulation, we would like to add edges to make it a triangulation while making sure that the maximal degree does not increase too much. We also have to ensure that the graph remains simple which may require us to add some additional vertices as well. Let f be a face of G with vertices $v_1, \ldots, v_k$. Suppose first that there are no edges between non-consecutive vertices of the face. In this case, we draw the edges in a zig-zag fashion, as in Fig. 5.3.

In the case where there are edges between non-consecutive vertices of the face exist, we draw a cycle $u_1, \ldots, u_k$ inside f. Then, we connect u_i to v_i and v_{i+1} for each $i < k$ and u_k to v_k and v_1. Finally, we triangulate the inner face created by the new cycle by zig-zagging as in the previous case (see Fig. 5.4).

Fig. 5.3 Adding diagonals to a face in a zigzag fashion

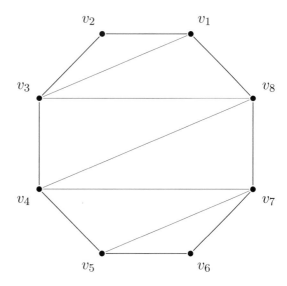

Fig. 5.4 Drawing an inner
cycle and triangulating the
new inner face

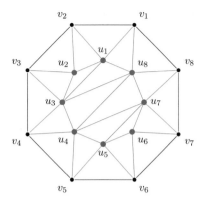

Since each vertex of the original graph is a member of at most D faces and for
each face at most 2 edges are added, the maximal degree of the resulting graph is
at most $3D$. Similarly, the number of vertices in the resulting graph is at most D
times the number of vertices in the original graph hence the probability of a random
vertex being a vertex of the original graph is at least D^{-1}. If this occurs then it is
straightforward to see that the existence of a subset of vertices B in the new graph
which satisfies the required conditions implies the existence of such a set in the old
graph, concluding our reduction to the triangulation case and finishing our proof. □

We are ready to deduce Theorem 5.8.

Proof of Theorem 5.8 Assume that G_n are finite planar maps with maximum degree
at most D such that $G_n \xrightarrow{\text{loc}} (U, \rho)$. If $\{G_n\}$ are not simple graphs we erase self-
loops and merge parallel edges into a single edge to obtain the sequence $\{G'_n\}$. It
is immediate that $G'_n \xrightarrow{\text{loc}} (U', \rho')$ where (U', ρ') is distributed as (U, ρ) after
removing from U all loops and merging parallel edges into a single edge. Since the
maximum degree is bounded, U' is recurrent if and only if U is recurrent. Thus we
may assume that G_n are simple graphs so the previous estimates may be used.

Denote by $\mathcal{A}_k$ the event

$$\mathcal{A}_k = \{\exists B \subseteq U, \ |B| \leq Ck, \ \mathcal{R}_{\text{eff}}(\rho \leftrightarrow V \setminus B) \geq c \log k\},$$

where $C = C(D) < \infty$ is the constant from Lemma 5.10. Therefore $\mathbb{P}(\mathcal{A}_k^c) \leq$
$c^{-1}k^{-\frac{1}{3}} \log(k)$. Looking at the sequence $\{\mathcal{A}_{2^j}\}_{j \geq 1}$, by Borel-Cantelli, almost surely
there exists j_0 such that for all $j \geq j_0$ the event $\mathcal{A}_{2^j}$ holds. Thus we have proved
the required assertion for k which is a power of 2. To prove this for all k sufficiently
large, let B_{2^j} be the set guaranteed to exist in the definition of $\mathcal{A}_{2^j}$, and take $B_k =
B_{2^j}$ for the unique j for which $2^j \leq k < 2^{j+1}$. It is immediate that these sets satisfy
the assertion of the theorem, concluding our proof.   □

5.4 Exercises

1. Let $G(n, p)$ be the random graph on n vertices drawn such that each of the $\binom{n}{2}$ possible edges appears with probability p independently of all other edges. Let $\lambda > 0$ be a constant, show that $G(n, \lambda/n)$ converges locally to a branching process with progeny distribution Poisson(λ).
2. For a graph G, let G^2 be the graph on the same vertex set as G so that vertices u, v form an edge if and only if the graph distance in G between u and v is at most 2. Show that if G has uniformly bounded degrees, then G is recurrent if and only if G^2 is recurrent.
3. Construct an example of a local limit (U, ρ) of finite planar graphs such that U is almost surely recurrent, but U^2 is almost surely transient.
4. Fix an integer $k \geq 1$. Construct an example of a sequence of finite simple planar maps G_n such that G_n converge locally to (U, ρ) with the property that $\mathbb{E}[\deg^k(\rho)] < \infty$ and U is almost surely transient.
5. (*) Suppose that G_n is a sequence of finite trees converging locally to (U, ρ). Show that U is almost surely recurrent. (Note that the degrees may be *unbounded*).

Chapter 6
Recurrence of Random Planar Maps

Our main goal in this chapter is to remove the bounded degrees assumption in Theorem 5.2 and replace it with the assumption that the degree of the root has an exponential tail.

Theorem 6.1 ([31]) *Let G_n be a sequence of (possibly random) planar graphs such that $G_n \xrightarrow{\text{loc}} (U, \rho)$ and there exist $C, c > 0$ such that $\mathbb{P}(\deg(\rho) \geq k) \leq Ce^{-ck}$ for every k. Then U is almost surely recurrent.*

As discussed in Sect. 1.2, the last theorem is immediately applicable in the setting of random planar maps. It is well known that the degree of the root in the UIPT and the UIPQ has an exponential tail. See [5, Lemma 4.1 and 4.2] or [26] for the UIPT and [8, Proposition 9] for the UIPQ.

Corollary 6.2 ([31]) *The UIPT/UIPQ are almost surely recurrent.*

6.1 Star-Tree Transform

We present here a transformation which transforms any planar map G to a planar map G^* with maximal degree of 4. We call this transformation $G \mapsto G^*$ the **star-tree transform**. Recall that a **balanced rooted tree** is a finite rooted tree in which every non-leaf vertex has precisely two children and the distance of the leaves from the root differs by at most 1. The transformation is performed as follows.

1. Subdivide each edge e by adding a new vertex w_e of degree two in the "middle". See Fig. 6.1b. Denote the resulting graph by G'.
2. For every vertex $v \in V(G)$, replace all edges incident to v in G' by a balanced binary tree rooted at v, whose leaves are the neighbors of v in G'. We perform this in a fashion which preserves the cyclic order of these neighbors and thus

© The Author(s) 2020
A. Nachmias, *Planar Maps, Random Walks and Circle Packing*, Lecture Notes in Mathematics 2243, https://doi.org/10.1007/978-3-030-27968-4_6

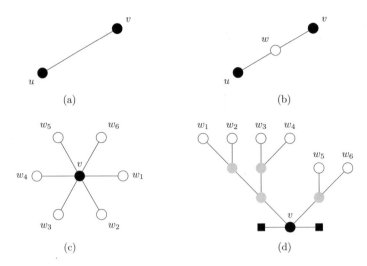

Fig. 6.1 The star-tree transform. (**a**) An original edge of G. (**b**) Subdividing an edge. (**c**) The "star" of a vertex in G'. (**d**) Transforming the star of v into a tree T_v

preserves planarity. Furthermore, add two extra vertices and attach them to the root. Denote this tree by T_v. See Fig. 6.1d.

Remark 6.3 The careful reader will notice that we have not specified precisely what is T_v (if $\deg_G(v)$ is not a power of 2 there may several balanced binary trees with $\deg(v)$ leaves) and in which way precisely we identify the leaves of T_v with the neighbors of v in G' (we may rotate the tree and get a different identification while still preserving planarity). This is a subtle yet important issue[1] and our convention is that the choice of tree and identification are performed uniformly at random from all the possible choices. This will be crucially used in Claim 6.13.

Lemma 6.4 *Let G be a planar map and G^* its star-tree transform. We set edge resistances on G^* by putting $R_e = 1/d_G(v)$, where v is the vertex of G for which $e \in T_v$ and $d_G(v)$ is the degree of v in G. If the network (G^*, R_e) is recurrent, then G is recurrent as well.*

Proof It is clear that from the point of view of recurrence versus transience, the two edges leading to the two "extra" neighbors of each root do not matter and can be removed. Hence for the rest of the proof we write T_v for the previously defined tree with these two edges removed. The purpose of these extra edges will become apparent later in the proof of Theorem 6.1.

Assume G is transient and let $a \in V(G)$ be some vertex. There is a flow θ from a to ∞ such that $\mathcal{E}(\theta) < \infty$. We will construct a flow θ^* on (G^*, R_e) from a to

---

[1] We thank Daniel Jerison for pointing this out to us.

∞ with finite energy, showing that (G^*, R_e) is transient, giving the theorem. First we define a flow θ' from a to infinity in G' in the natural manner: for each edge $e = (x, y)$ of G we set $\theta'(x, w_e) = \theta'(w_e, y) = \theta(x, y)$. Obviously $\mathcal{E}(\theta') = 2\mathcal{E}(\theta)$.

Next we provide some notation. We denote by A the set of vertices that were added to form G' in the first step of the star-tree transform, that is, the white vertices in Fig. 6.1. Each vertex $w \in A$ is a leaf of precisely two trees T_u and T_v, where $\{u, v\}$ was the edge of G that w divided. We call u and v the **tree roots** of w. We denote by B the set of vertices that were added to G^* in the second step of the star-tree transform, that is, the gray vertices in Fig. 6.1d. The vertices of $V(G)$ are the black discs in Fig. 6.1. Each vertex of $x \in V(G) \cup B$ is a member of a single tree T_v; we call v the **tree root** of x. Lastly, for any $x \in V(G) \cup B$ we denote by $C_x \subset A$ the set of leaves of T_v, where v is the tree root of x, for which the path from the leaf to the root of T_v goes through x; in other words, C_x is the set of leaves of T_v which are the "descendants" of x. If $x \in A$, then we set $C_x = \{x\}$.

To define θ^*, let $e = (x, y)$ be an edge of T_v. Assume that x is closer to the root of T_v than y in graph distance. We set

$$\theta^*(e) = \sum_{w \in C_y} \theta'(v, w).$$

The construction of θ^* is depicted in Fig. 6.2.

We will now show that $\mathcal{E}(\theta^*) \leq 2\mathcal{E}(\theta')$ where the energy of θ^* is taken in the network (G^*, R_e). Let $v \in V(G)$ and write h for the height of T_v, that is, h is the maximal graph distance from a leaf of T_v to its root. Note that since the tree

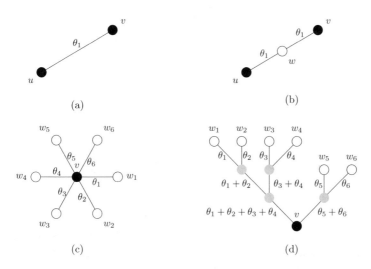

Fig. 6.2 The construction of the flow θ^* from θ. (**a**) An original edge of G which has flow θ_1. (**b**) The flow passes through the divided edge. (**c**) The flow going out from a vertex of G in G'. (**d**) The division of the flow in T_v

is balanced, the distances from the leaves to the root vary by at most 1. Let $e = (x, y)$ be an edge of T_v and assume that x is closer than y to the root of T_v. By the construction of θ^*, the contribution of e to $\mathcal{E}(\theta^*)$ is

$$R_e\theta^*(e)^2 = \frac{1}{d_G(v)}\left(\sum_{w\in C_y}\theta'(v,w)\right)^2.$$

If the graph distance of y from the root is $\ell \in \{1,\ldots,h\}$, then $|C_y| \leq 2^{h-\ell}$. Hence by Cauchy-Schwarz

$$R_e\theta^*(e)^2 \leq \frac{2^{h-\ell}}{d_G(v)}\sum_{w\in C_y}\theta'(v,w)^2.$$

Summing over all edges in T_v at distance ℓ from the root, we go over each leaf of T_v precisely once. Thus,

$$\sum_{\substack{e=(x,y)\in T_v \\ d_{G^*}(y,v)=\ell}} R_e\theta^*(e)^2 \leq \frac{2^{h-\ell}}{d_G(v)}\sum_{w\in C_v}\theta'(v,w)^2.$$

We now sum over all edges in T_v by summing over $\ell \in \{1,\ldots,h\}$. We get

$$\sum_{e\in T_v} R_e\theta^*(e)^2 \leq \frac{2^h}{d_G(v)}\sum_{w\in C_v}\theta'(v,w)^2 \leq 2\sum_{w\in C_v}\theta'(v,w)^2,$$

since $h \leq \log_2(d_G(v)) + 1$. Lastly, we sum this over all $v \in V(G)$ to obtain that

$$\mathcal{E}(\theta^*) \leq 2\mathcal{E}(\theta') = 4\mathcal{E}(\theta),$$

concluding our proof. $\square$

6.2   Stationary Random Graphs and Markings

Stationary Random Graphs

Recall that Theorem 6.1 and the entire setup of Chap. 5 is adapted to the case when G_n is itself random. The reason is that in Definition 5.1 we consider the graph distance ball $B_{G_n}(\rho_n, r)$ as a random variable in the probability space $(\mathcal{G}_\bullet, d_{\mathrm{loc}})$, where ρ_n conditioned on G_n is a uniformly chosen random vertex.

Let us emphasize that this is **not** the same as drawing a sample of $\{G_n\}$ and claiming that almost surely $G_n \xrightarrow{\text{loc}} (U, \rho)$. For example, let G_n be a path of length n with probability $1/2$ and an $n \times n$ square grid with probability $1/2$, independently for all n. In this case $G_n \xrightarrow{\text{loc}} (U, \rho)$ where $U = \mathbb{Z}$ with probability $1/2$ and $U = \mathbb{Z}^2$ with probability $1/2$, however, almost surely on the sequence $\{G_n\}$, the local limit of G_n does not exist.

In many cases it is useful to take a random root drawn from the stationary distribution on G_n, that is, the probability distribution on vertices giving each vertex v probability $\deg_{G_n}(v)/2|E(G_n)|$. In a similar fashion to Definition 5.1, we define this type of local convergence.

Definition 6.5 Let $\{G_n\}$ be a sequence of (possibly random) finite graphs with non-empty sets of edges. We say that $G_n \xrightarrow[\pi]{\text{loc}} (U, \rho)$ where (U, ρ) is a random rooted graph, if for every integer $r \geq 1$,

$$B_{G_n}(\rho_n, r) \xrightarrow{d} B_U(\rho, r),$$

where ρ_n is a randomly chosen vertex from G_n with distribution proportional to the vertex degrees. We call such a limit a **stationary local limit**.

Let us remark that $G_n \xrightarrow{\text{loc}} (U, \rho)$ does not imply that $G_n \xrightarrow[\pi]{\text{loc}} (U', \rho')$ for some (U', ρ'). Indeed, let G_n be a path of length n attached to a complete graph on $\sqrt{n}$ vertices. Then the local limit of G_n is $\mathbb{Z}$, however the limit according to a stationary random root does not exist.

The reason for taking the $\xrightarrow[\pi]{\text{loc}}$ limit rather than the uniform limit as before is that the random walk on the limit (U, ρ) starting from ρ is then stationary.

Claim 6.6 Assume that $G_n \xrightarrow[\pi]{\text{loc}} (U, \rho)$. Conditioned on (U, ρ), let X_1 be a uniformly chosen neighbor of ρ. Then (U, X_1) is equal in law to (U, ρ). Similarly, if $\{X_n\}_{n \geq 0}$ is the simple random walk on (U, ρ), then for each $n \geq 0$ the law of (U, X_n) coincides with the law of (U, ρ).

Proof If H is a finite graph and v is a vertex chosen with probability proportional to its degree, then it is immediate that a uniformly chosen random neighbor of v is distributed according to the stationary distribution. Thus for any fixed $r > 0$ the ball $B_{G_n}(\rho_n, r)$ has the same distribution as $B_{G_n}(X_1, r)$ where ρ_n is drawn from the stationary distribution on G_n and X_1 is a uniform neighbor of ρ_n. The claim follows now by definition. $\square$

Definition 6.7 A random rooted graph (G, ρ) is called a **stationary random graph** if (G, X_1) has the same distribution as (G, ρ), where the vertex X_1 is a uniform neighbor of ρ conditioned on (G, ρ).

We would like to develop a simple abstract framework that will allow us to comfortably move from $\xrightarrow{\text{loc}}$ convergence to $\xrightarrow[\pi]{\text{loc}}$ convergence and vice versa. This is straightforward when $\{G_n\}$ are a sequence of *deterministic* graphs with uniformly bounded average degree but is less obvious when G_n themselves are random. For this we need to *degree bias* our random graphs.

Definition 6.8 Denote by $\mathbb{P}$ the law of a random rooted graph (G, ρ) and assume that $\mathbb{E}\deg(\rho) \in (0, \infty)$. The probability measure μ on $(\mathcal{G}_\bullet, d_{\text{loc}})$ defined by

$$\mu(\mathcal{A}) := \frac{1}{\mathbb{E}\deg(\rho)} \sum_{k \geq 1} k \, \mathbb{P}(\mathcal{A} \cap \{\deg(\rho) = k\}),$$

for any event $\mathcal{A} \subset (\mathcal{G}_\bullet, d_{\text{loc}})$ is called the **degree biasing** of $\mathbb{P}$. Similarly, if we assume that almost surely ρ is not an isolated vertex, then the probability measure ν defined by

$$\nu(\mathcal{A}) = \frac{1}{\mathbb{E}[\deg(\rho)^{-1}]} \sum_{k \geq 1} \frac{\mathbb{P}(\mathcal{A} \cap \{\deg(\rho) = k\})}{k},$$

is called the **degree unbiasing** of $\mathbb{P}$.

Lemma 6.9 *Assume that (G, ρ) is a random rooted graph such that G is almost surely finite, that the distribution of ρ given G is uniform and that $\mathbb{E}\deg(\rho) \in (0, \infty)$. Then the degree biasing of (G, ρ) is a stationary random graph.*

Conversely, assume that (G^π, ρ^π) is a stationary random graph such that G^π is almost surely finite and has no isolated vertices. Then its degree unbiasing (G, ρ) is such that G is almost surely finite and ρ conditioned on G is uniformly distributed.

Proof We will prove only the first statement and the second is similar. Denote by (G^π, ρ^π) a random variable drawn according to the degree biasing of (G, ρ). Let H be a fixed finite graph and denote by $\deg_H(v)$ the degree of a vertex v in H. By definition we have that

$$\mathbb{P}((G^\pi, \rho^\pi) = (H, v)) = \frac{\deg_H(v) \cdot \mathbb{P}((G, \rho) = (H, v))}{\mathbb{E}\deg(\rho)}. \tag{6.1}$$

Let X_1 be a uniformly chosen neighbor of ρ^π. Then by (6.1)

$$\mathbb{P}((G^\pi, X_1) = (H, u)) = \sum_{v : \{u,v\} \in E(H)} \frac{\mathbb{P}((G^\pi, \rho^\pi) = (H, v))}{\deg_H(v)} = \frac{\sum_{v : \{u,v\} \in E(H)} \mathbb{P}((G, \rho) = (H, v))}{\mathbb{E}\deg(\rho)}.$$

Since ρ is uniformly distributed given G, the quantity $\mathbb{P}((G, \rho) = (H, v))$ is the same for all v. So

$$\mathbb{P}((G^\pi, X_1) = (H, u)) = \frac{\deg_H(u) \mathbb{P}((G, \rho) = (H, u))}{\mathbb{E}\deg(\rho)}$$

so by (6.1) the required assertion follows.   □

Corollary 6.10 *Assume that $\{G_n\}$ is a sequence of random graphs that are almost surely finite and that $\mathbb{E}\deg(\rho_n) \in (0, \infty)$ where ρ_n is a uniformly chosen vertex of G_n. Let (G_n^π, ρ_n^π) be the degree biasing of (G_n, ρ_n). Assume that $G_n \overset{loc}{\longrightarrow} (U, \rho)$ and that $\mathbb{E}\deg(\rho) < \infty$ and that $\mathbb{E}\deg(\rho_n) \to \mathbb{E}\deg(\rho)$. Then $G_n^\pi \overset{loc}{\underset{\pi}{\longrightarrow}} (U^\pi, \rho^\pi)$ where (U^π, ρ^π) is the degree biasing of (U, ρ). Furthermore, (U, ρ) and (U^π, ρ^π) are absolutely continuous with respect to each other.*

Conversely, assume that $\{G_n^\pi\}$ is a sequence of random graphs that are almost surely finite and have no isolated vertices. Denote by ρ_n^π a random vertex of G_n^π drawn with probability proportional to the vertex degrees and by (G_n, ρ_n) the degree unbiasing of (G_n^π, ρ_n^π). If $G_n^\pi \overset{loc}{\underset{\pi}{\longrightarrow}} (U^\pi, \rho^\pi)$, then $G_n \overset{loc}{\longrightarrow} (U, \rho)$ where (U, ρ) is the degree unbiasing of (U^π, ρ^π). Furthermore, (U, ρ) and (U^π, ρ^π) are absolutely continuous with respect to each other.

Proof We start by proving the first assertion. Let (H, v) be a finite rooted graph and $r > 0$ a fixed integer. Then by Definition 6.8

$$\mathbb{P}(B_{G_n^\pi}(\rho_n^\pi, r) = (H, v)) = \frac{\deg_H(v)\mathbb{P}(B_{G_n}(\rho_n, r) = (H, v))}{\mathbb{E}\deg(\rho_n)}.$$

Since $G_n \overset{loc}{\longrightarrow} (U, \rho)$ and $\mathbb{E}\deg(\rho_n) \to \mathbb{E}\deg(\rho)$ we obtain that

$$\lim_{n\to\infty} \mathbb{P}(B_{G_n^\pi}(\rho_n^\pi, r) = (H, v)) = \frac{\deg_H(v)\mathbb{P}(B_U(\rho, r) = (H, v))}{\mathbb{E}\deg(\rho)} = \mathbb{P}(B_{U^\pi}(\rho^\pi, r) = (H, v)),$$

where the last equality is also by Definition 6.8. The absolute continuity of (U, ρ) and (U^π, ρ^π) follows immediately from the definition.

The second statement follows by the same proof. Note that $\mathbb{E}[\deg(\rho_n^\pi)^{-1}] \to \mathbb{E}[\deg(\rho^\pi)^{-1}]$ by definition since $B_{G_n^\pi}(\rho_n^\pi, 1)$ converges in distribution to $B_{U^\pi}(\rho^\pi, 1)$ and the function $f((G, \rho)) = \deg(\rho)^{-1}$ is a bounded continuous function on $\mathcal{G}_\bullet$.   □

We end this subsection by addressing the somewhat technical issue of verifying the condition $\mathbb{E}\deg(\rho_n) \to \mathbb{E}\deg(\rho)$ in Corollary 6.10. It is not guaranteed just by requiring $\sup_n \mathbb{E}\deg(\rho_n) < \infty$ as can be seen in the example of a path of length n where we choose $\sqrt{n}$ arbitrary vertices and add $\sqrt{n}$ loops to each one; in this example $\deg(\rho) = 2$ almost surely, and $\mathbb{E}\deg(\rho_n) = 4 + o(1)$. However, we now show that it is always possible to "truncate" the finite graphs G_n by removing edges touching vertices of large degrees so that the limit is unchanged and the average degrees converge to the expected degree of the limit. Given a finite graph G and an integer $k \geq 1$ we denote by $G \wedge k$ the graph obtained from G by erasing all the edges touching vertices of degree at least k. We note that even when G is connected, $G \wedge k$ may be disconnected and may have isolated vertices. As we defined in Sect. 5.1, by $(G \wedge k, \rho)$ we mean $(G \wedge k[\rho], \rho)$ where $G \wedge k[\rho]$ is the connected component of

ρ in $G \wedge k$, hence it is a member of $\mathcal{G}_\bullet$ even when it is disconnected. All statements in this chapter, most importantly Corollary 6.10, do not assume the graphs involved are connected.

Lemma 6.11 *Let $\{G_n\}$ be a sequence of random finite graphs such that $G_n \overset{\text{loc}}{\longrightarrow} (U, \rho)$ and $\mathbb{E}\deg(\rho) < \infty$. Then there exists a sequence $k(n) \to \infty$ such that*

$$G_n \wedge k(n) \overset{\text{loc}}{\longrightarrow} (U, \rho).$$

Furthermore, if we set $G'_n = G_n \wedge k(n)$, then

$$\mathbb{E}\deg_{G'_n}(\rho_n) \to \mathbb{E}\deg(\rho),$$

where ρ_n is a uniformly chosen vertex of G'_n.

Proof We first show that for *any* sequence $k(n) \to \infty$ we have that $G_n \wedge k(n) \overset{\text{loc}}{\longrightarrow} (U, \rho)$. Indeed, since $G_n \overset{\text{loc}}{\longrightarrow} (U, \rho)$ we have that for any fixed integer $r \geq 1$

$$\mathbb{P}\left(\max\left\{\deg(v) : v \in B_{G_n}(\rho_n, r)\right\} \geq k(n)\right) \to 0.$$

If $\max\{\deg(v) : v \in B_{G_n}(\rho_n, r+1)\} < k(n)$, then $B_{G_n}(\rho_n, r) = B_{G_n \wedge k(n)}(\rho_n, r)$. Since G_n and $G_n \wedge k(n)$ have the same set of vertices we deduce that for any fixed $r \geq 1$ and any rooted graph (H, v)

$$\mathbb{P}\big(B_{G_n \wedge k(n)}(\rho_n, r) = (H, v)\big) \to \mathbb{P}(B_U(\rho, r) = (H, v)).$$

Secondly, since $\deg(\rho_n)$ converges in distribution to $\deg(\rho)$ we have that there exists a sequence $k(n) \to \infty$ such that $\mathbb{E}[\deg(\rho_n) \wedge k(n)] \to \mathbb{E}\deg(\rho)$. Indeed, by dominated convergence we have that $\mathbb{E}[\deg(\rho) \wedge k] \to_{k\to\infty} \mathbb{E}\deg(\rho)$. Furthermore, for any fixed k the function $f((G, \rho)) = \deg(\rho) \wedge k$ is a bounded and continuous on $\mathcal{G}_\bullet$, thus $\mathbb{E}[\deg(\rho_n) \wedge k] \to_{n\to\infty} \mathbb{E}[\deg(\rho) \wedge k]$. Hence for any $\varepsilon > 0$ there exist k and N such that for all $n \geq N$ we have that $|\mathbb{E}[\deg(\rho_n) \wedge k] - \mathbb{E}\deg(\rho)| \leq \varepsilon$. It is an exercise that this implies the existence of $k(n)$.

Lastly, $\limsup \mathbb{E}\deg_{G'_n}(\rho_n) \leq \mathbb{E}\deg(\rho)$ since $\deg_{G'_n}(\rho_n) \leq \deg_{G_n}(\rho_n) \wedge k(n)$. We also have that $\deg_{G'_n}(\rho_n) \overset{d}{\longrightarrow} \deg(\rho)$, hence by Fatou's lemma $\liminf \mathbb{E}\deg_{G'_n}(\rho_n) \geq \mathbb{E}\deg(\rho)$, and hence the second assertions follows. $\square$

Markings

Given a locally convergent sequence of (possibly random) graphs G_n, we wish to apply the star-tree transform on them to create a sequence G_n^* and take its local limit of that while "remembering", in light of Lemma 6.4, the original degrees of G_n. The approach is a rather straightforward extension of the abstract setting of Sect. 5.1, see also [2]. We consider the space of triples (G, ρ, M) where $G = (V, E)$ is a graph, $\rho \in V$ is a vertex and $M : E \to \mathbb{R}$ is a function assigning real values to the edges. We endow the space with a metric by setting the distance between (G_1, ρ_1, M_1) and (G_2, ρ_2, M_2) to be 2^{-R} where R is the maximal value such that there exists a rooted graph isomorphism φ between $B_{G_1}(\rho_1, R)$ and $B_{G_2}(\rho_2, R)$ such that $|M_1(e) - M_2(\varphi(e))| \le R^{-1}$ for all edges $e \in E(G)$ both of whose end points are in $B_{G_1}(\rho_1, R)$. It is easy to check that this space is again a Polish space, so again we may define convergence in distribution of random variables taking values in this space.

We say that such a random triplet (U, ρ, M) is **stationary** if conditioned on (U, ρ, M) a uniformly chosen random neighbor X_1 of ρ satisfies that (U, ρ, M) has the same law as (U, X_1, M) in the space of isomorphism classes of rooted graphs with markings (that is, rooted isomorphisms that preserve the markings). Given a marking M we extend it to $M : E(U) \cup V(U) \to \mathbb{R}$ by setting $M(v) = \max_{e: v \in e} M(e)$ for any $v \in V(U)$. We say that (U, ρ, M) has an **exponential tail** if for some $A < \infty$ and $\beta > 0$ we have that $\mathbb{P}(M(\rho) \ge s) \le A e^{-\beta s}$ for all $s \ge 0$.

In the following lemma we consider a stationary triplet (U, ρ, M) that has an exponential tail and compare the hitting probabilities of certain sets when we endow the graphs with two sets of edge resistances: the first are the usual unit resistances, and in the second we may change the edge resistances arbitrarily but only on edges with high M values. We tailored the lemma this way in order to show that (G^*, R_e) from Lemma 6.4 is recurrent.

Lemma 6.12 *Let (U, ρ, M) be a stationary, bounded degree rooted random graph with markings which has an exponential tail. Conditioned on (U, ρ, M) and given some finite set $B \subset U$, let $\mathbf{P}_\rho$ denote the unit-resistance random walk on U starting from ρ and let $\mathbf{P}'_\rho$ denote the random walk on U with edge resistances R'_e satisfying that $R'_e = 1$ whenever $M(e) \le 21\beta^{-1} \log |B|$. Then almost surely on (U, ρ, M) there exists $K < \infty$ such that for any finite subset $B \subset U$ with $|B| \ge K$ we have*

$$\left| \mathbf{P}_\rho(\tau_{U \setminus B} < \tau_\rho^+) - \mathbf{P}'_\rho(\tau_{U \setminus B} < \tau_\rho^+) \right| \le \frac{1}{|B|} .$$

Proof For every pair of integers $T, s \ge 1$ we set

$$A_{T,s} = \left\{ \mathbf{P}_\rho(\exists t < T : M(X_t) \ge s) \le T^3 e^{-\beta s/2} \right\} .$$

Since (U, ρ, M) is stationary and has an exponential tail for any $t \ge 0$ we have

$$\mathbb{E}\left[\mathbf{P}_\rho(M(X_t) \ge s) \right] \le A e^{-\beta s} ,$$

hence by the union bound

$$\mathbb{E}\left[\mathbf{P}_\rho(\exists t < T : M(X_t) \ge s) \right] \le A T e^{-\beta s} .$$

Thus by Markov's inequality

$$\mathbb{P}\left(\mathcal{A}_{T,s}^c\right) \le AT^{-2}e^{-\beta s/2}.$$

By Borel-Cantelli we deduce that almost surely $\mathcal{A}_{T,s}$ occurs for all but finitely many pairs T, s. Conditioned on (U, ρ, M), we may consider only finite subsets $B \subset U$ which contain ρ, since otherwise both probabilities in the statement of the lemma are 1. Let B be such a subset. By the commute time identity Lemma 2.26, and since the maximum degree of U is bounded,

$$\mathbf{E}_\rho(\tau_{U\setminus B}) \le C\mathcal{R}_{\text{eff}}(\rho \leftrightarrow U \setminus B)|B| \le C|B|^2,$$

for some constant $C > 0$. The last inequality is since the resistance is bounded by $|B|$ since there is a path of length at most $|B|$ from ρ to $U \setminus B$. By Markov's inequality,

$$\mathbf{P}_\rho(\tau_{U\setminus B} \ge T) \le \frac{C|B|^2}{T}.$$

Write $S = \{v \in U : M(v) \ge s\}$. For every T, s for which $\mathcal{A}_{T,s}$ occurs we have

$$\mathbf{P}_\rho\left(\tau_S < \tau_{\{\rho\}\cup U\setminus B}^+\right) \le \mathbf{P}_\rho(\tau_{U\setminus B} \ge T) + \mathbf{P}_\rho(\exists t < T : M(X_t) \ge s) \le \frac{C|B|^2}{T} + T^3e^{-\beta s/2}.$$

We now choose $T = 2C|B|^3$ and $s = 21\beta^{-1} \log |B|$ so that the right hand side of the last inequality is at most $|B|^{-1}$ when $|B|$ is sufficiently large. It is clear that we can couple two random walks starting from ρ, one walking on U with unit resistances and the other on (U, R_e), so that they remain together until they visit a vertex of S. Hence, when $|B|$ is large enough so that the chosen T, s are such that $\mathcal{A}_{T,s}$ holds we deduce from the last inequality that with probability at least $1 - |B|^{-1}$ the simple random walk on U visits $\{\rho\} \cup U \setminus B$ before visiting S, concluding our proof.   □

6.3 Proof of Theorem 6.1

We now proceed to wrapping up the proof of Theorem 6.1. Recall that we have a sequence of finite planar graphs $\{G_n\}$ such that $G_n \xrightarrow{\text{loc}} (U, \rho)$ and with $\mathbb{P}(\deg(\rho) \ge k) \le Ce^{-ck}$. Our goal is to prove that (U, ρ) is almost surely recurrent.

Let us explain how we use Lemma 6.11 and Corollary 6.10 to truncate and degree bias G_n and (U, ρ) so that we may assume without loss of generality that $G_n \xrightarrow{\text{loc}}$ (U, ρ). Indeed, if it does not hold that $\mathbb{E}\deg(\rho_n) \to \mathbb{E}\deg(\rho)$ we consider $G_n \wedge k(n)$ of Lemma 6.11 which has the same limit (U, ρ). Since $k(n) \to \infty$ the graphs $G_n \wedge k(n)$ have non-empty set of edges (we assume that G_n have non-empty sets of edges otherwise (U, ρ) is an isolated vertex), and thus we may apply Corollary 6.10 and deduce that the degree biasing $(G_n \wedge k(n), \rho_n)$ converges to the degree biasing of (U, ρ) which is absolutely continuous with respect to (U, ρ), and in particular, it is recurrent almost surely if and only if (U, ρ) is. We also erase from $G_n \wedge k(n)$ all isolated vertices that may have been created in the truncation, since these are drawn with probability 0 after the degree bias. This will be important for us later when we unbias the graphs. Lastly, it is an easy computation using Definition 6.8 that we still have $\mathbb{P}(\deg(\rho) \geq k) \leq Ce^{-ck}$ (possibly for some other positive constants C, c). Thus, from now on we assume without loss of generality that $G_n \xrightarrow[\pi]{\text{loc}} (U, \rho)$ and that $\deg(\rho)$ has an exponential tail and that G_n have no isolated vertices almost surely.

Recall now the definitions and notations of Sect. 6.1. Consider the star-tree transform G_n^* of G_n and let ρ_n^* be a random vertex of T_{ρ_n} drawn according to the stationary distribution of T_{ρ_n}. Similarly, conditioned on (U, ρ), let U^* be the star-tree transform of U and ρ^* be a random vertex of T_ρ drawn according to the stationary distribution of T_ρ. Furthermore, we put markings on G_n^* and U^* by marking each edge e of G_n^* or U^* with $\deg(v)$ whenever e is in the tree T_v and $\deg(v)$ is the degree of v in G_n or U, respectively. Denote these markings by M_n and M, respectively.

Claim 6.13 We have that (G_n^*, ρ_n^*, M_n) for each n and (U^*, ρ^*, M) are stationary, and,

$$(G_n^*, \rho_n^*, M_n) \xrightarrow{d} (U^*, \rho^*, M).$$

Proof Since for any fixed integer $r > 0$, the laws of $B_{G_n^*}(\rho_n^*, r)$ and $B_{U^*}(\rho^*, r)$ are determined by $B_{G_n}(\rho_n, r)$ and $B_U(\rho, r)$, respectively, see Remark 6.3. We obtain that

$$(G_n^*, \rho_n^*, M_n) \xrightarrow{d} (U^*, \rho^*, M).$$

Secondly, it is immediate to check that for each $v \in G_n$ we have that the number of edges in T_v is precisely $2 \deg_{G_n}(v)$. This is the reason why we added the two "extra" neighbors to the root of T_v in the star tree transform described in Sect. 6.1. Thus, conditioned on G_n, for any $x \in G_n^*$ such that $x \in T_v$ for some $v \in G_n$ we have that

$$\mathbb{P}(\rho_n^* = x \mid G_n) = \frac{\deg_{G_n}(v)}{2|E(G_n)|} \cdot \frac{\deg_{T_v}(x)}{2|E(T_v)|} = \frac{\deg_{T_v}(x)}{2|E(G_n^*)|},$$

or in other words, (G_n^*, ρ_n^*, M_n) is a stationary random graph and since it converges to (U^*, ρ^*, M), the latter is also stationary.   □

Lemma 6.14 *The triplet* (U^*, ρ^*, M) *has an exponential tail.*

Proof We observe that $M(\rho^*) = \deg(v)$ where v is either ρ or one of its neighbors. Hence it suffices to show that if (U, ρ) is a stationary local limit such that $\deg(\rho)$ has an exponential tail, then the random variable $D(\rho) = \max_{v:\{\rho,v\}\in E(U)} \deg(v)$ has an exponential tail. We have

$$\mathbb{P}(D(\rho) \geq k) \leq \mathbb{P}(\deg(\rho) \geq k) + \mathbb{P}(\deg(\rho) \leq k \text{ and } D(\rho) \geq k). \tag{6.2}$$

The probability of the first term on the right hand side decays exponentially in k due to our assumption on (U, ρ). Conditioned on (U, ρ), let X_1 be a uniformly chosen random neighbor of ρ. Then clearly

$$\mathbb{P}(\deg(X_1) \geq k \mid \deg(\rho) \leq k \text{ and } D(\rho) \geq k) \geq k^{-1}.$$

However, by stationarity $\mathbb{P}(\deg(X_1) \geq k) = \mathbb{P}(\deg(\rho) \geq k)$, which decays exponentially. We conclude that the second term on the right hand side of (6.2) decays exponentially as well. □

Consider the stationary random graph (U^*, ρ^*, M). By Lemma 6.14 it has an exponential tail. Consider the edge resistances

$$R_e^{\text{unit}} \equiv 1, \qquad R_e^{\text{mark}} = \frac{1}{M(e)}.$$

In view of Lemma 6.4, it suffices to show that the network (U^*, R^{mark}) is almost surely recurrent, for then it will follow that U is almost surely recurrent. To prove the former, we apply the second assertion of Corollary 6.10 which allows us to assume without loss of generality that (U^*, ρ^*) is a local limit of finite planar maps (rather than a stationary local limit). In the beginning of the proof we assumed that almost surely G_n have no isolated vertices (they were erased after the degree biasing), hence the same holds for G_n^* and we may use Corollary 6.10. Since (U^*, ρ^*) is now a local limit of finite planar maps with degrees bounded by 4 we may apply Theorem 5.8 to obtain an almost sure constant $c > 0$ and a sequence of sets $B_k \subset U^*$ such that

1. $ck \leq |B_k| \leq c^{-1}k$, and
2. $\mathcal{R}_{\text{eff}}(\rho^* \leftrightarrow U^* \setminus B_k \;; \{R_e^{\text{unit}}\}) \geq c \log k$,

where we added to the conclusion of Theorem 5.8 that $B_k \geq ck$ since adding vertices to B_k makes the lower bound on the resistance even better.

We now define one extra set of edge resistances on U^* which will allow us to interpolate between the edge resistances R^{unit} and R^{mark}. For each integer $k \geq 1$ we define

$$R_e^{\text{mid}} = \begin{cases} 1 & M(e) \leq C \log k, \\ M^{-1}(e) & \text{otherwise}, \end{cases}$$

where $C > 0$ is some large constant that will be chosen later. We will use $\mathbf{P}$, $\mathbf{P}^{\mathrm{mark}}$ and $\mathbf{P}^{\mathrm{mid}}$ to denote the probability measures, conditioned on (U^*, ρ^*, M), of random walks on U^* with edge resistances $\{R_e^{\mathrm{unit}}\}$, $\{R_e^{\mathrm{mark}}\}$ and $\{R_e^{\mathrm{mid}}\}$, respectively.

Lemma 6.15 *For some other constant* $c > 0$ *we have*

$$\mathcal{R}_{\mathrm{eff}}(\rho^* \leftrightarrow U^* \setminus B_k \; ; \{R_e^{\mathrm{mid}}\}) \geq c \log k \,.$$

Proof We may assume k is large enough so that $M(e) \leq C \log k$ for every edge e incident to ρ^*. By Claim 2.22 we have

$$\mathcal{R}_{\mathrm{eff}}(\rho^* \leftrightarrow U^* \setminus B_k \; ; \{R_e^{\mathrm{unit}}\}) \leq \frac{1}{\mathbf{P}_{\rho^*}(\tau_{U^* \setminus B_k} < \tau_{\rho^*}^+)} \,,$$

hence

$$\mathbf{P}_{\rho^*}(\tau_{U^* \setminus B_k} < \tau_{\rho^*}^+) \leq \frac{1}{c \log k} \,,$$

by our assumption on B_k above. By Lemma 6.12 it follows that

$$\mathbf{P}_{\rho^*}^{\mathrm{mid}}(\tau_{U^* \setminus B_k} < \tau_{\rho^*}^+) \leq \frac{2}{c \log k} \,,$$

when k is large enough and the constant $C > 0$ in the definition of $\{R_e^{\mathrm{mid}}\}$ is chosen large enough with respect to β. Using Claim 2.22 again and the fact that U^* has degrees bounded by 4 concludes the proof. $\square$

We need yet another easy general fact about electric networks.

Claim 6.16 Consider a finite network G in which all resistances are bounded above by 1. Then for any integer $m \geq 1$ and any two vertices $a \neq z$ we have

$$\mathcal{R}_{\mathrm{eff}}(B_G(a, m) \leftrightarrow z) \geq \mathcal{R}_{\mathrm{eff}}(a \leftrightarrow z) - m \,.$$

Proof Let θ^m be the unit current flow from $B(a, m)$ to z. For a vertex $v \in B(a, m)$ denote

$$\alpha_v = \sum_{u \notin B(a,m): u \sim v} \theta^m(vu)$$

so that $\alpha_v \geq 0$ for all $v \in B(a, m)$ and $\sum_{v \in B(a,m)} \alpha_v = 1$. For a vertex $v \in B(a, m)$ let $\theta^{a,v}$ be a unit flow putting flow 1 on some shortest path from a to v in $B(a, m)$. Set

$$\theta = \sum_{v \in B(a,m)} \alpha_v (\theta^m + \theta^{a,v}) \,.$$

By Thomson's principle (Theorem 2.28), Jensen's inequality and since $\sum_v \alpha_v = 1$ we have

$$\mathcal{R}_{\text{eff}}(a \leftrightarrow z) \leq \mathcal{E}(\theta) = \mathcal{E}(\theta^m) + \sum_e r_e \Big[\sum_{v \in B(a,m)} \alpha_v \theta^{a,v}(e) \Big]^2 \leq \mathcal{E}(\theta^m) + \sum_{v \in B(a,m)} \alpha_v \sum_e r_e \left(\theta^{a,v}(e) \right)^2$$

$$\leq \mathcal{E}(\theta^m) + \sum_{v \in B(a,m)} \alpha_v \cdot m = \mathcal{R}_{\text{eff}}(B(a,m) \leftrightarrow z) + m \,. \qquad \square$$

We are finally ready to conclude the proof of the main theorem of this chapter.

Proof of Theorem 6.1 By Lemma 6.15 and Claim 6.16 we have that the sets B_k obtained earlier satisfy that for any $m \geq 0$

$$\mathcal{R}_{\text{eff}}(B_{U^*}(\rho^*, m) \leftrightarrow U^* \setminus B_k \; ; \{R_e^{\text{mid}}\}) \geq c \log k - m \,.$$

Moreover, for every edge e,

$$R_e^{\text{mark}} \geq \frac{R_e^{\text{mid}}}{C \log k} \,,$$

hence

$$\mathcal{R}_{\text{eff}}(B_{U^*}(\rho^*, m) \leftrightarrow U^* \setminus B_k \; ; \{R_e^{\text{mark}}\}) \geq c/C - m/C \log k \,.$$

By taking $k \to \infty$ we deduce that there exists $c > 0$ such that for any $m \geq 1$

$$\mathcal{R}_{\text{eff}}(B_{U^*}(\rho^*, m) \leftrightarrow \infty; \{R_e^{\text{mark}}\}) \geq c \,.$$

Consider the current unit flow from ρ^* to ∞ in $(U^*, \{R_e^{\text{mark}}\})$. If this flow had finite energy, then for any $\varepsilon > 0$ there would exists $m \geq 1$ such that $\mathcal{R}_{\text{eff}}(B_{U^*}(\rho^*, m) \leftrightarrow \infty; \{R_e^{\text{mark}}\}) \leq \varepsilon$, which is a contradiction to the above. Hence

$$\mathcal{R}_{\text{eff}}(\rho^* \leftrightarrow \infty; \{R_e^{\text{mark}}\}) = \infty \,,$$

that is, $(U^*, \{R_e^{\text{mark}}\})$ is almost surely recurrent. The theorem now follows by Lemma 6.4. $\qquad \square$

Chapter 7
Uniform Spanning Trees of Planar Graphs

7.1 Introduction

Let G be a finite connected graph. A **spanning tree** T of G is a connected subgraph of G that contains no cycles and such that every vertex of G is incident to at least one edge of T. The set of spanning trees of a given finite connected graph is obviously finite and hence we may draw one uniformly at random. This random tree is called the **uniform spanning tree** (UST) of G. This model was first studied by Kirchhoff [49] who gave a formula for the number of spanning trees of a given graph and provided a beautiful connection with the theory of electric networks. In particular, he showed that the probability that a given edge $\{x, y\}$ of G is contained in the UST equals $\mathcal{R}_{\text{eff}}(x \leftrightarrow y; G)$; we prove this fundamental formula in Sect. 7.2 (see Theorem 7.2).

Is there a natural way of defining a UST probability measure on an infinite connected graph? It will soon become clear that we have set the framework already in Sect. 2.3 to answer this question positively. Let $G = (V, E)$ be an infinite connected graph and assume that $\{G_n\}$ is a finite exhaustion of G as defined in Sect. 2.5. That is, $\{G_n\}$ is a sequence of finite graphs, $G_n \subset G_{n+1}$ for all n, and $\cup G_n = G$. Russell Lyons conjectured that the UST probability measure on G_n converges weakly to some probability measure on subsets of E and in his pioneering work Pemantle [68] showed that it is indeed the case.

More precisely, denote by $\mathcal{T}_n$ a UST of G_n, then it is shown in [68] that for any two finite subset of edges A, B of G the limit

$$\lim_{n \to \infty} \mathbf{P}(A \subset \mathcal{T}_n , \; B \cap \mathcal{T}_n = \emptyset), \tag{7.1}$$

exists and does not depend on the exhaustion $\{G_n\}$. The proof is a consequence of Rayleigh's monotonicity (Corollary 2.29) and will be presented in Sect. 7.3. This together with Kolmogorov's extension theorem [24, Theorem A.3.1] implies that there exists a unique probability measure on infinite subsets of E for which a sample

A. Nachmias, *Planar Maps, Random Walks and Circle Packing*, Lecture Notes in Mathematics 2243, https://doi.org/10.1007/978-3-030-27968-4_7

of $\mathfrak{F}$ satisfies

$$\mathbf{P}(A \subset \mathfrak{F}, \; B \cap \mathfrak{F} = \emptyset) = \lim_{n \to \infty} \mathbf{P}(A \subset \mathcal{T}_n, \; B \cap \mathcal{T}_n = \emptyset),$$

for any two finite subsets of edges A and B of G. Thus, the law of $\mathfrak{F}$ is determined and we denote it by μ^F. The superscript F stands for *free* and will be explained momentarily. Let us explore some properties of μ^F that are immediate from its definition.

Since every vertex of G is touched by at least one edge of $\mathcal{T}_n$ with probability 1 when n is large enough (so that G_n contains the vertex), we learn that the edges of $\mathfrak{F}$ almost surely touch every vertex of G, that is, $\mathfrak{F}$ is almost surely *spanning*. Similarly, the probability that the edges of a given cycle in G are contained in $\mathcal{T}_n$ (once n is large enough so that G_n contains the cycle) is 0. Since G has countably many cycles we deduce that almost surely there are no cycles in $\mathfrak{F}$. By a similar reasoning we deduce that almost surely any connected component of $\mathfrak{F}$ is infinite. However, a moment's reflection shows that this kind of reasoning cannot be used to determine that $\mathfrak{F}$ is almost surely connected.

It turns out, perhaps surprisingly, that $\mathfrak{F}$ need not be connected almost surely. A remarkable result of Pemantle [68] shows that a sample of μ^F on $\mathbb{Z}^d$ is almost surely connected when $d = 1, 2, 3, 4$ and almost surely disconnected when $d \geq 5$. Since it may be the case that a sample of μ^F is disconnected with positive probability, we call μ^F the **free uniform spanning forest** (rather than tree) of G, denoted henceforth FUSF_G. The term *free* corresponds to the fact that we have not imposed any boundary conditions when taking a limit. It will be very useful to take other boundary conditions, such as the *wired* boundary condition, see Sect. 7.3. The seminal paper of Benjamini et al. [12] explores many properties of these infinite random trees (properties such as number of components and connectivity in particular, size of the trees, recurrence or transience of the trees and many others) on various underlying graphs with an emphasis on Cayley graphs. We refer the reader to [12] and to [61, Chapters 4 and 10] for a comprehensive treatment.

The question of connectivity of the FUSF is therefore fundamental and unfortunately it is not even known that connectivity is an event of probability 0 or 1 on any graph G, see [12, Question 15.7]. In [44] the circle packing theorem (Theorem 3.5) is used to prove that FUSF_G is almost surely connected when G is a bounded degree proper planar map, answering a question of [12, Question 15.2]. Our goal in this chapter is to present a proof for a specific case where G is a bounded degree, transient, one-ended planar triangulations. Even though this is a particular case of a general theorem, the argument we present here contains most of the key ideas. We refer the interested reader to [44] for the general statement.

Theorem 7.1 ([44]) *Let G be a simple, bounded degree, transient, one-ended planar triangulation. Then FUSF_G is almost surely connected.*

The rest of this chapter is organized as follows. In Sect. 7.2 we discuss two basic properties of USTs on finite graphs. Namely, Kirchhoff's effective resistance formula mentioned earlier and the spatial Markov property for the UST. In Sect. 7.3 we prove Pemantle's [68] result (7.1) showing that FUSF_G exists. We will also define there the *wired uniform spanning forest* which is obtained by taking a limit of the UST probability measures over exhaustions with wired boundary. We will also need some fairly basic notions of electric networks on infinite graphs that we have not discussed in Sect. 2.5. Next, in Sect. 7.4 we will restrict to the setting of planar graph and employ planar duality to obtain an extremely useful connection between the free and wired spanning forests which will be useful later. Using these tools we have collected we will prove Theorem 7.1 in Sect. 7.5.

7.2 Basic Properties of the UST

Kirchhoff's Effective Resistance Formula

Theorem 7.2 (Kirchoff [49]) *Let G be a finite connected graph and denote by $\mathcal{T}$ a uniformly drawn spanning tree of G. Then for any edge $e = (x, y)$ we have*

$$\mathbf{P}(e \in \mathcal{T}) = \mathcal{R}_{\mathrm{eff}}(x \leftrightarrow y).$$

Proof Let $a \neq z$ be two distinct vertices of G (later we will take $a = x$ and $z = y$) and note that any spanning tree of G contains precisely one path connecting a and z. Thus, a uniformly drawn spanning tree induces a random path from a to z. By Claim 2.46 we obtain a unit flow θ from a to z. To be concrete, for each edge e we have that $\theta(\vec{e})$ is the probability that the random path from a to z traverses $\vec{e}$ minus the probability that it traverses $\overleftarrow{e}$. We will now show that θ satisfies the cycle law (see Claim 2.14), so it is in fact the unit current flow (see Definition 2.19).

Let $\vec{e}_1, \ldots, \vec{e}_m$ be a directed cycle in G. Our goal is to show that

$$\sum_{i=1}^{m} \theta(\vec{e}_i) = 0. \tag{7.2}$$

Denote by $T(G)$ the set of spanning trees of G. Expanding the sum on the left hand side with the definition of θ we get that it equals

$$|T(G)|^{-1} \sum_{t \in T(G)} \sum_{i=1}^{m} f_i^+(t) - |T(G)|^{-1} \sum_{t \in T(G)} \sum_{j=1}^{m} f_j^-(t),$$

where $f_i^+(t)$ equals 1 if the unique path from a to z in t traverses $\vec{e}_i$ and 0 otherwise, and similarly, $f_j^-(t)$ equals 1 if this path traverses $\overleftarrow{e}_j$ and 0 otherwise.

For $1 \leq i \leq m$ we denote by T_i^+ the set of pairs (t, i) for which $f_i^+(t) = 1$. Similarly define T_j^- as the set of pairs (t, j) for which $f_j^-(t) = 1$. To prove (7.2) it suffices to show that

$$| \uplus_{i\in\{1,...m\}} T_i^+ | = | \uplus_{j\in\{1,...m\}} T_j^- | .$$

Let $(t, i) \in T_i^+$. The graph $t \setminus \{e_i\}$ has two connected components. Let $\vec{e}_j$ be the first edge after $\vec{e}_i$, in the order of the cycle $\vec{e}_1, \ldots, \vec{e}_m$, that is incident to both connected components and consider the spanning tree $t' = t \cup \{e_j\} \setminus \{e_i\}$. Note that the unique path in t' from a to z traverses $\overleftarrow{e}_j$, so $(t', j) \in T_j^-$. This procedure defines a bijection from $\uplus_i T_i^+$ to $\uplus_j T_j^-$. Indeed, given (t', j) from before, we can erase e_j and go on the cycle in the opposite order until we reach e_i which has to be the first edge incident to the two connected components of $t' \setminus \{e_j\}$. This shows (7.2) and concludes the proof.   □

Spatial Markov Property of the UST

We would like to study the UST probability measure conditioned on the event that some edges are present in the UST and others not. It turns out that sampling from this conditional distribution amounts to drawing a UST on a modified graph.

Let $G = (V, E)$ be a finite connected graph and let A and B be two disjoint subsets of edges. We write $(G - B)/A$ for the graph obtained from G by erasing the edges of B and contracting the edges of A. We identify the edges of $(G - B)/A$ with the edges $E \setminus B$. Denote by $\mathcal{T}_G$ and $\mathcal{T}_{(G-B)/A}$ a UST on G and $(G - B)/A$, respectively, and assume that

$$\mathbf{P}(A \subset \mathcal{T}_G , B \cap \mathcal{T}_G = \emptyset) > 0 .$$

This assumption is equivalent to $G - B$ being connected and that A contains no cycles.

Then, conditioned on the event that $\mathcal{T}_G$ contains the edges A and does not contain any edge of B the distribution of $\mathcal{T}_G$ is equal to the union of A with $\mathcal{T}_{(G-B)/A}$. In other words, for a set $\mathscr{A}$ of spanning trees of G we have that

$$\mathbf{P}(\mathcal{T}_G \in \mathscr{A} \mid A \subset \mathcal{T}_G , B \cap \mathcal{T}_G = \emptyset) = \mathbf{P}(A \cup \mathcal{T}_{(G-B)/A} \in \mathscr{A}) . \tag{7.3}$$

The proof of (7.3) follows immediately from the observation that the set of spanning trees of G not containing any edge of B is simply the set of spanning trees of $G - B$. Similarly, the set of spanning trees of G containing all the edges of A is simply the union of A to each spanning tree of G/A, and (7.3) follows.

7.3   Limits over Exhaustions: The Free and Wired USF

Let G be an infinite connected graph and let $\{G_n\}$ be a finite exhaustion of it. In this section we will show that (7.1) holds and that the UST measures with *wired* boundary conditions also converge. Let us first explain the latter. Denote by G_n^* the graph obtained from G by identifying the infinite set of vertices $G \setminus G_n$ to a single vertex z_n and erasing the loops at z_n formed by this identification. We say that $\{G_n^*\}$ is a **wired finite exhaustion** of G.

Theorem 7.3 (Pemantle [68]) *Let G be an infinite connected graph, $\{G_n\}$ a finite exhaustion and $\{G_n^*\}$ the corresponding wired finite exhaustion. Denote by $\mathcal{T}_n$ and $\mathcal{T}_n^*$ USTs on G_n and G_n^*, respectively. Then for any two finite disjoint subsets $A, B \subset E(G)$ of edges of G we have that the limits*

$$\lim_{n \to \infty} \mathbf{P}(A \subset \mathcal{T}_n \, , \, B \cap \mathcal{T}_n = \emptyset) \, ,$$

and

$$\lim_{n \to \infty} \mathbf{P}(A \subset \mathcal{T}_n^* \, , \, B \cap \mathcal{T}_n^* = \emptyset) \, ,$$

exist and do not depend on the exhaustion $\{G_n\}$.

We postpone the proof for a little longer and first discuss some of its implications. As mentioned earlier, Theorem 7.3 together with Kolmogorov's extension theorem [24, Theorem A.3.1] implies that there exists two probability measures μ^F and μ^W on infinite subsets of the edges of E arising as the unique limits of the laws $\mathcal{T}_n$ and $\mathcal{T}_n^*$. That is, the samples $\mathfrak{F}^f$ and $\mathfrak{F}^w$ of μ^F and μ^W satisfy

$$\mathbf{P}(A \subset \mathfrak{F}^f \, , \, B \cap \mathfrak{F}^f = \emptyset) = \lim_{n \to \infty} \mathbf{P}(A \subset \mathcal{T}_n \, , \, B \cap \mathcal{T}_n = \emptyset) \, ,$$

and

$$\mathbf{P}(A \subset \mathfrak{F}^w \, , \, B \cap \mathfrak{F}^w = \emptyset) = \lim_{n \to \infty} \mathbf{P}(A \subset \mathcal{T}_n^* \, , \, B \cap \mathcal{T}_n^* = \emptyset) \, .$$

We call μ^F and μ^W the **free uniform spanning forest** and the **wired uniform spanning forest** and denote them by FUSF_G and WUSF_G respectively. We have seen earlier (one paragraph below (7.1)) that both $\mathfrak{F}^f$ and $\mathfrak{F}^w$ are almost surely spanning forests, that is, spanning graphs of G with no cycles and that every connected component of them is infinite. Thus μ^F and μ^W are supported on what are known as **essential spanning forests** of G, that is, spanning forests of G in which every component is infinite.

Are the probability measures FUSF_G and WUSF_G equal? Not necessarily. It is easy to see that on the infinite path $\mathbb{Z}$ the $\mathsf{WUSF}_\mathbb{Z}$ and the $\mathsf{FUSF}_\mathbb{Z}$ are equal and are the entire graph $\mathbb{Z}$ with probability 1. Conversely, it is not very difficult to see that they are different on a 3-regular tree, see exercise 1 of this chapter. Pemantle [68] has shown that $\mathsf{FUSF}_{\mathbb{Z}^d} = \mathsf{WUSF}_{\mathbb{Z}^d}$ for any $d \geq 1$ and a very useful criterion for determining whether there is equality was developed in [12]. We refer the reader to [61, Chapter 10] for further reading.

Before presenting the proof of Theorem 7.3 let us make a few short observations regarding the effective resistance between two vertices in an infinite graph, extending what we proved in Sect. 2.5.

Effective Resistance in Infinite Networks

Let G be an infinite connected graph. We have seen in Sect. 2.5 that for any vertex v the electric resistance $\mathcal{R}_{\mathrm{eff}}(v \leftrightarrow \infty)$ from v to ∞ is well defined as the limit of $\mathcal{R}_{\mathrm{eff}}(a \leftrightarrow z_n; G_n^*)$ where $\{G_n^*\}$ is a wired finite exhaustion and z_n is the vertex resulting in the identification of the vertices $G \setminus G_n$.

To define the electric resistance between two vertices v, u of an infinite graph, one has to take exhaustions and specify boundary conditions since the limits may differ depending on them.

Claim 7.4 Let G be an infinite connected graph, $\{G_n\}$ a finite exhaustion and $\{G_n^*\}$ a wired finite exhaustion. Then for any two vertices u, v of G we have that the limits

$$\mathcal{R}_{\mathrm{eff}}^F(u \leftrightarrow v; G) := \lim_n \mathcal{R}_{\mathrm{eff}}(u \leftrightarrow v; G_n),$$

and

$$\mathcal{R}_{\mathrm{eff}}^W(u \leftrightarrow v; G) := \lim_n \mathcal{R}_{\mathrm{eff}}(u \leftrightarrow v; G_n^*),$$

exist and do not depend on the exhaustion $\{G_n\}$.

Proof For the first limit we note that by Rayleigh's monotonicity (Corollary 2.29), the sequence $\mathcal{R}_{\mathrm{eff}}(u \leftrightarrow v; G_n)$ is non-increasing and non-negative since $G_n \subset G_{n+1}$, hence it converges. A sandwiching argument as in the proof of Claim 7.4 shows that the limit does not depend on the exhaustion $\{G_n\}$.

For the second limit, since G_n can be obtained by gluing vertices of G_{n+1} we deduce by Corollary 2.30 that the sequence $\mathcal{R}_{\mathrm{eff}}(u \leftrightarrow v; G_n^*)$ is non-decreasing and bounded (by the graph distance in G between u and v for instance), hence it converges. The limit does not depend on the exhaustion by an identical sandwiching argument.   □

We call $\mathcal{R}_{\mathrm{eff}}^F(u \leftrightarrow v; G)$ and $\mathcal{R}_{\mathrm{eff}}^W(u \leftrightarrow v; G)$ the **free effective resistance** and **wired effective resistance** between u and v respectively.

Proof of Theorem 7.3

We will prove the assertion regarding the first limit; the second is almost identical.
Write $A = \{e_1, \ldots, e_k\}$ and $e_i = (x_i, y_i)$ for each $1 \leq i \leq k$. Assume without loss
of generality that G_n contains A for all n. As before, denote by $\mathcal{T}_n$ a UST of G_n.
By (7.3) and Theorem 7.2 we have that

$$\mathbf{P}(A \subset \mathcal{T}_n) = \prod_{i=1}^{k} \mathbf{P}(e_i \in \mathcal{T}_n \mid e_j \in \mathcal{T}_n \quad \forall j < i) = \prod_{i=1}^{k} \mathcal{R}_{\mathrm{eff}}(x_i \leftrightarrow y_i; G_n/\{e_1, \ldots, e_{i-1}\}).$$

Note that $\{G_n/\{e_1, \ldots, e_{i-1}\}\}$ is a finite exhaustion of the infinite graph
$G/\{e_1, \ldots, e_{i-1}\}$ and so by Claim 7.4 we obtain that the limit

$$\lim_n \mathbf{P}(A \subset \mathcal{T}_n) = \prod_{i=1}^{k} \mathcal{R}_{\mathrm{eff}}(x_i \leftrightarrow y_i; G/\{e_1, \ldots, e_{i-1}\}),$$

exists and does not depend on the exhaustion.

Since we know this limit exists for all finite edge sets A, it follows by the
inclusion-exclusion formula that $\mathbf{P}(A \subset \mathcal{T}_n, B \cap \mathcal{T}_n = \emptyset)$ converges for any finite
sets A, B, concluding our proof.   □

It is now quite pleasant to see that the symbiotic relationship between electric
network and UST theories continues to flourish in the infinite setting. Indeed, by
combining Theorems 7.3 and Claim 7.4 we obtain the extension of Kirchhoff's
formula for infinite connected graphs.

Theorem 7.5 *Let G be an infinite connected graph and denote by $\mathfrak{F}^F$ and $\mathfrak{F}^W$ a
sample from* FUSF_G *and* WUSF_G *respectively. Then for any edge $e = (x, y)$ of G
we have that*

$$\mathbf{P}(e \in \mathfrak{F}^F) = \mathcal{R}_{\mathrm{eff}}^F(x \leftrightarrow y; G),$$

and

$$\mathbf{P}(e \in \mathfrak{F}^W) = \mathcal{R}_{\mathrm{eff}}^W(x \leftrightarrow y; G).$$

7.4 Planar Duality

When G is planar there is a very useful relationship between FUSF_G and WUSF_G.
Recall that given a planar map G, the **dual graph** of G is the graph $G^\dagger$ whose vertex
set is the set of faces of G and two faces are adjacent in $G^\dagger$ if they share an edge in
G. Thus, $G^\dagger$ is locally-finite if and only if every face of G has finitely many edges.

To each edge $e \in E(G)$ corresponds a dual edge $e^{\dagger} \in E(G^{\dagger})$ which is the pair of faces of G incident to e; this is clearly a one-to-one correspondence.

When G is a finite planar graph, this correspondence induces a one-to-one correspondence between the set of spanning trees of G and the set of spanning trees of $G^{\dagger}$. Given a spanning tree of t of G we slightly abuse the notation and write $t^{\dagger}$ for the set of edges $\{e^{\dagger} : e \in G \setminus t\}$, that is

$$e \in t \iff e^{\dagger} \notin t^{\dagger}.$$

If $t^{\dagger}$ has a cycle, then t is disconnected. Furthermore, if there is a vertex $G^{\dagger}$ not incident to any edge of $t^{\dagger}$, then all the edges of the corresponding face in G are present in t hence t contains a cycle. We deduce that if t is a spanning tree of G, then $t^{\dagger}$ is a spanning tree of $G^{\dagger}$. The converse also holds since $(t^{\dagger})^{\dagger} = t$ and $(G^{\dagger})^{\dagger} = G$.

Now assume that G is an infinite planar maps such that $G^{\dagger}$ is locally finite. Given an essential spanning forest $\mathfrak{F}$ of G we similarly define $\mathfrak{F}^{\dagger}$ as the set of edges $\{e^{\dagger} : e \in G \setminus \mathfrak{F}\}$. A similar argument shows that $\mathfrak{F}^{\dagger}$ is an essential spanning forest of $G^{\dagger}$. This raises the natural question: when $\mathfrak{F}$ is a sample of FUSF_G, what is the law of $\mathfrak{F}^{\dagger}$? The answer in general is an object known as the *transboundary uniform spanning forest* [44, Proposition 5.1]. However, when G is additionally assumed to be one-ended (in particular, in the setting of Theorem 7.1) it turns out that $\mathfrak{F}^{\dagger}$ is distributed as $\mathsf{WUSF}_{G^{\dagger}}$:

Proposition 7.6 *Let G be an infinite, one-ended planar map with a locally finite dual $G^{\dagger}$ and let $\mathfrak{F}$ be a sample of FUSF_G. Then the law of $\mathfrak{F}^{\dagger}$ is $\mathsf{WUSF}_{G^{\dagger}}$.*

Proof Let G_n be a finite exhaustion of G. Let F_n be a finite exhaustion $G^{\dagger}$ defined by letting $f \in F_n$ if and only if every vertex of f in G belongs to G_n. Then $G_n^{\dagger}$ is obtained from $G^{\dagger}$ by contracting $G^{\dagger} \setminus F_n$ into a single vertex which corresponds to the outer face of G_n. Thus, $G_n^{\dagger}$ is a wired exhaustion of $G^{\dagger}$ and the statement follows.   □

We use to obtain an important criterion of connectivity of FUSF_G in the planar case.

Proposition 7.7 *Let G be an infinite, one-ended planar map with a locally finite dual $G^{\dagger}$. Then a sample of FUSF_G is connected almost surely if and only if each component of a sample of WUSF_G is one-ended almost surely.*

Proof By Proposition 7.6 it suffices to show that if $\mathfrak{F}$ is an essential spanning forest of G, then $\mathfrak{F}$ is connected if and only if every component of $\mathfrak{F}^{\dagger}$ is one-ended. Indeed, if $\mathfrak{F}$ is disconnected, then the boundary of a connected component of $\mathfrak{F}$ induces an bi-infinite path in $\mathfrak{F}^{\dagger}$. Conversely, if $\mathfrak{F}^{\dagger}$ contains a bi-infinite path, then by the Jordan curve theorem $\mathfrak{F}$ is disconnected.   □

7.5 Connectivity of the Free Forest

Last Note on Infinite Networks

We make two more useful and natural definitions. Given two disjoint finite sets A and B in an infinite connected graph G we define the free and wired effective resistance between them $\mathcal{R}^W_{\text{eff}}(A \leftrightarrow B; G)$ and $\mathcal{R}^F_{\text{eff}}(A \leftrightarrow B; G)$ as the free and wired effective resistance between a and b in the graph obtained from G by identifying A and B to the vertices a and b.

Lastly, given a graph G, a wired finite exhaustion $\{G^*_n\}$ of G and two disjoint finite sets A and B we define

$$\mathcal{R}_{\text{eff}}(A \leftrightarrow B \cup \{\infty\}; G) := \lim_{n \to \infty} \mathcal{R}_{\text{eff}}(A \leftrightarrow B \cup \{z_n\}; G^*_n), \qquad (7.4)$$

where the last limit exists since the sequence is non-increasing from n that is large enough so that G_n contains A and B. In the proof of Theorem 7.1 we will require the following estimate.

Lemma 7.8 *Let A and B be two finite sets of vertices in an infinite connected graph G. Then*

$$\mathcal{R}^W_{\text{eff}}(A \leftrightarrow B; G) \leq 3 \max \left[\mathcal{R}_{\text{eff}}(A \leftrightarrow B \cup \{\infty\}; G), \mathcal{R}_{\text{eff}}(B \leftrightarrow A \cup \{\infty\}; G) \right].$$

Proof For any three distinct vertices u, v, w in a finite network we have by the union bound that $\mathbf{P}_u(\tau_{\{v,w\}} < \tau^+_u) \leq \mathbf{P}_u(\tau_v < \tau^+_u) + \mathbf{P}_u(\tau_w < \tau^+_u)$. Hence by Claim 2.22 we get that

$$\mathcal{R}_{\text{eff}}(u \leftrightarrow \{v, w\})^{-1} \leq \mathcal{R}_{\text{eff}}(u \leftrightarrow v)^{-1} + \mathcal{R}_{\text{eff}}(u \leftrightarrow w)^{-1}.$$

Let $\{G^*_n\}$ be a wired finite exhaustion of G and assume without loss of generality that A and B are contained in G^*_n for all n. Then by the previous estimate

$$\mathcal{R}_{\text{eff}}(A \leftrightarrow B \cup \{z_n\}; G^*_n)^{-1} \leq \mathcal{R}_{\text{eff}}(A \leftrightarrow B; G^*_n)^{-1} + \mathcal{R}_{\text{eff}}(A \leftrightarrow z_n; G^*_n)^{-1}.$$

Denote by M the maximum in the statement of the lemma and take $n \to \infty$ in the last inequality. We obtain that

$$M^{-1} \leq \mathcal{R}_{\text{eff}}(A \leftrightarrow B \cup \{\infty\}; G)^{-1} \leq \mathcal{R}^W_{\text{eff}}(A \leftrightarrow B; G)^{-1} + \mathcal{R}_{\text{eff}}(A \leftrightarrow \infty; G)^{-1}.$$

Rearranging gives that

$$\mathcal{R}_{\text{eff}}(A \leftrightarrow \infty; G) \leq \frac{M \mathcal{R}^W_{\text{eff}}(A \leftrightarrow B; G)}{\mathcal{R}^W_{\text{eff}}(A \leftrightarrow B; G) - M}.$$

By symmetry, the same inequality holds when we replace the roles of A and B. We put this together with the triangle inequality for effective resistances (2.9) and get that

$$\mathcal{R}_{\text{eff}}^W(A \leftrightarrow B; G) \leq \mathcal{R}_{\text{eff}}(A \leftrightarrow \infty; G) + \mathcal{R}_{\text{eff}}(B \leftrightarrow \infty; G) \leq \frac{2M\mathcal{R}_{\text{eff}}^W(A \leftrightarrow B; G)}{\mathcal{R}_{\text{eff}}^W(A \leftrightarrow B; G) - M},$$

which by rearranging gives the desired inequality.   □

Method of Random Sets

We present the following weakening of the method of random paths as in Sect. 2.6. Let μ be the law of a random subset W of vertices of G. Define the *energy* of μ as

$$\mathcal{E}(\mu) = \sum_{v \in V} \mu(v \in W)^2.$$

Lemma 7.9 (Method of Random Sets) *Let A, B be two disjoint finite sets of vertices in an infinite graph G. Let W be a random subset of vertices of G and denote by μ its law. Assume that the subgraph of G induced by W almost surely contains a simple path starting at A that is either infinite or finite and ends at B. Then*

$$\mathcal{R}_{\text{eff}}(A \leftrightarrow B \cup \{\infty\}; G) \leq \mathcal{E}(\mu). \tag{7.5}$$

Proof Given W let γ be a simple path, contained in W, connecting A to B or an infinite path starting at A. We choose γ according to some prescribed lexicographical ordering. Then, letting ν be the law of γ,

$$\mathcal{E}(\nu) \leq \sum_{\vec{e} \in E} \nu(\vec{e} \in \gamma)^2,$$

where by $\vec{e} \in \gamma$ we mean that the directed edge $\vec{e}$ is traversed (in its direction) by γ, and by $\mathcal{E}(\nu)$ we mean the energy of the flow induced by γ, as in Claim 2.46.

Let γ' be an independent random path having the same law as γ. Then the sum above is precisely the expected number of directed edges traversed both by γ and γ'. Since these are simple paths, they each contain at most one directed edge emanating from each vertex $v \in W$. Thus, the expected number of directed edges used by both paths is at most the number of vertices used by both paths. Hence,

$$\mathcal{E}(\nu) \leq \sum_{v \in V(G)} \nu(v \in \gamma)^2 \leq \sum_{v \in V(G)} \mu(v \in W)^2 = \mathcal{E}(\mu),$$

and the proof is concluded by Thomson's principle (Theorem 2.28).   □

Proof of Theorem 7.1

In Theorem 7.1 we assume that $G = (V, E)$ is a bounded-degree, one-ended triangulation. Hence $G^\dagger$ is a bounded degree (in fact, 3-regular), one-ended and transient planar map with faces of uniformly bounded size. We leave this verification as an exercise for the reader. To avoid carrying the $\dagger$ symbol around, and with a slight abuse of notation, let $G = (V, E)$ be a graph satisfying these assumptions on $G^\dagger$, that is, we assume that G is a one-ended, transient, infinite planar map with bounded degrees and face sizes. We will prove under these assumptions that every component of WUSF_G is one ended almost surely which implies Theorem 7.1 by Proposition 7.7.

Let T be the bounded-degree one-ended triangulation obtained from G by adding a vertex inside each face of G and connecting it by edges to the vertices of that face according to their cyclic ordering. By Theorem 4.4 there exists a circle packing of T in the unit disc $\mathbb{U}$. We identify the vertices of T as the vertices $V(G)$ and faces $F(G)$ of G, and denote this circle packing as $P = \{P(v) : v \in V(G)\} \cup \{P(f) : f \in F(G)\}$.

Given $z \in \mathbb{U}$ and $r' \geq r > 0$ denote by $A_z(r, r')$ the annulus $\{w \in \mathbb{C} : r \leq |w - z| \leq r'\}$.

Definition 7.10 Write $V_z(r, r')$ for the set of vertices v of G such that either

- $P(v) \cap A_z(r, r') \neq \emptyset$, or
- $P(v) \subset \{w \in \mathbb{C} : |w| \leq r\}$ and there is a face f of G with $v \in f$ and $P(f) \cap A_z(r, r') \neq \emptyset$.

We emphasize that $V_z(r, r')$ contains only vertices of G; no vertices of T that correspond to faces of G belong to it.

Lemma 7.11 *There exists a constant* $C < \infty$ *depending only on the maximal degree such that for any* $z \in \mathbb{U}$ *and any positive integer n satisfying* $|z| \geq 1 - C^{-n}$ *the sets*

$$V_z(C^{-i}, 2C^{-i}) \qquad 1 \leq i \leq n,$$

are disjoint.

Proof By the Ring Lemma (Lemma 4.2) there exists a constant $B < \infty$ such that for any $C > 1$, any z satisfying $z \geq 1 - C^{-n}$ and any $1 \leq i \leq n$, if a circle of P intersects $A_z(C^{-i}, 2C^{-i})$ or is tangent to a circle that intersects $A_z(C^{-i}, 2C^{-i})$, then its radius is at most BC^{-i}. Hence, this set of circles is contained in the disc of radius $(2 + 4B)C^{-i}$ around z. Furthermore, since $|z| \geq 1 - C^{-n}$, by the Ring Lemma again there exists $b > 0$ such that any such circle must be of distance at least bC^{-i} from z. Hence, any fixed $C > \frac{4+4B}{b}$ satisfies the assertion of the lemma. $\square$

Lemma 7.12 *Let $z \in \mathbb{U}$ and $r > 0$. Let U be a uniform random variable in $[1, 2]$ and denote by μ_r the law of the random set $V_z(Ur, Ur)$ (as defined in Definition 7.10). Then there exists a constant $C < \infty$ depending only on the maximal degree such that*

$$\mathcal{E}(\mu_r) \leq C .$$

Proof For each vertex v, the event $v \in V_z(Ur, Ur)$ implies that the circle $\{w \in \mathbb{C} : |w - z| = Ur\}$ intersects the circle $P(v)$ or intersects $P(f)$ for some face f incident to v. The union of $P(v)$ and $P(f)$ over all such faces f is contained in the Euclidean ball around the center of $P(v)$ of radius $r(v) + 2 \max_{f:v \in f} r(f)$. Since T has finite maximal degree we have that $r(f) \leq Cr(v)$ for all f with $v \in f$ where $C < \infty$ depends only on the maximal degree by the Ring Lemma (Lemma 4.2). Hence,

$$\mu_r(v \in V_z(Ur, Ur)) \leq \frac{1}{r} \min \left(2r(v) + 4 \max_{f:f \ni v} r(f), r \right) \leq \frac{C}{r} \min\{r(v), r\}.$$

$$\tag{7.6}$$

We claim that

$$\sum_{v \in V_z(r, 2r)} \min\{r(v), r\}^2 \leq 16r^2.$$

$$\tag{7.7}$$

Indeed, consider a vertex $v \in V_z(r, 2r)$ for which the corresponding circle $P(v)$ has radius larger than r. By Definition 7.10 this circle must intersect $\{w \in \mathbb{C} : |w - z| \leq 2r\}$. We replace each such $P(v)$ with a circle of radius r that is contained in the original circle and intersects $\{w \in \mathbb{C} : |w - z| \leq 2r\}$. The circles in this new set still have disjoint interiors and are contained in $\{w \in \mathbb{C} : |w - z| \leq 4r\}$. Therefore their area is at most $\pi 16r^2$ and (7.7) follows. The proof of lemma is now concluded by combining (7.6) and (7.7).   □

Proof of Theorem 7.1 Let $\mathfrak{F}$ be a sample of WUSF_G and given an edge $e = (x, y)$ we define $\mathscr{A}^e$ to be the event that x and y are in two distinct infinite connected components of $\mathfrak{F} \setminus \{e\}$. It is clear that every component of $\mathfrak{F}$ is one-ended almost surely if and only if

$$\mathbf{P}(e \in \mathfrak{F}, \mathscr{A}^e) = 0 \tag{7.8}$$

for every edge e of G. Consider the triangulation T described above Definition 7.10 and its circle packing P in $\mathbb{U}$. By choosing the proper Möbius transformation we may assume that the tangency point between $P(x)$ and $P(y)$ is the origin, and that the centers of $P(x)$ and $P(y)$ lie on the negative and positive real axis, respectively.

Fix now an arbitrary $\varepsilon > 0$ and let V_ε be all the vertices of G such that the center $z(v)$ of $P(v)$ satisfies $|z(v)| \leq 1 - \varepsilon$. Denote by $\mathscr{B}_\varepsilon^e$ the event that every connected

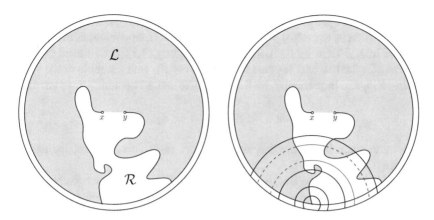

Fig. 7.1 Illustration of the proof. Left: On the event $\mathscr{A}_\varepsilon^e$, the paths η^x and η^y split V_ε into two pieces, $\mathcal{L}$ and $\mathcal{R}$. Right: We define a random set containing a path (solid blue) from η^x to $\eta^y \cup \{\infty\}$ in $G \setminus K_c$ using a random circle (dashed blue). Here we see two examples, one in which the path ends at η^y, and the other in which the path ends at the boundary (i.e., at infinity)

component of $\mathfrak{F} \setminus \{e\}$ intersects $V \setminus V_\varepsilon$. Note that $\mathcal{A}^e \subset \cap_{\varepsilon > 0} \mathscr{B}_\varepsilon^e$ but this containment is strict since it is possible that $e \notin \mathfrak{F}$ and x is connected to y in $\mathfrak{F}$ inside V_ε.

Assume that $\mathscr{B}_\varepsilon^e$ holds. Let η^x be the rightmost path in $\mathfrak{F} \setminus \{e\}$ from x to $V \setminus V_\varepsilon$ when looking at x from y, and let η^y be the leftmost path in $\mathfrak{F} \setminus \{e\}$ from y to $V \setminus V_\varepsilon$ when looking at y from x. As mentioned above, the paths η_x and η_y are not necessarily disjoint. Nonetheless, concatenating the reversal of η^x with e and η^y separates V_ε into two sets of vertices, $\mathcal{L}$ and $\mathcal{R}$, which are to the left and right of e (when viewed from x to y) respectively. See Fig. 7.1 for an illustration of the case when η_x and η_y are disjoint (when they are not, $\mathcal{R}$ is a "bubble" separated from $V \setminus V_\varepsilon$).

On the event $\mathscr{B}_\varepsilon^e$, let K be the set of edges that are either incident to a vertex in $\mathcal{L}$ or belong to the path $\eta_x \cup \eta_y$, and set $K = E$ off of this event. Note that the edges of K do not touch the vertices of $\mathcal{R}$. The condition that η^x and η^y are the rightmost and leftmost paths to $V \setminus V_\varepsilon$ from x and y is equivalent to the condition that K does not contain any open path from x to $V \setminus V_\varepsilon$ other than η^x, and does not contain any open path from y to $V \setminus V_\varepsilon$ other than η^y. We note that K can be explored algorithmically, without querying the status of any edge in $E \setminus K$, by performing a right-directed depth-first search of x's component in $\mathfrak{F}$ and a left-directed depth-first search of y's component in $\mathfrak{F}$, stopping each search when it first leaves V_ε.

Denote by $\mathscr{A}_\varepsilon^e$ the event that η_x and η_y are disjoint, or equivalently, that K does not contain an open path from x to y (and in particular, no path starting at η_x and ending at η_y). The event $\mathscr{A}_\varepsilon^e$ is measurable with respect to the random set K and $\mathscr{A}^e = \cap_{\varepsilon > 0} \mathscr{A}_\varepsilon^e$. Hence

$$\mathbf{P}(e \in \mathfrak{F}, \, \mathscr{A}^e) \leq \mathbf{P}(e \in \mathfrak{F} \mid \mathscr{A}_\varepsilon^e) = \mathbf{E}[\mathbf{P}(e \in \mathfrak{F} \mid \mathscr{A}_\varepsilon^e, K)]. \tag{7.9}$$

Denote by K_o the open edge of K (that is, the edge of K in $\mathfrak{F}$) and by K_c the closed edges of K (that is, the edges of K not belonging to $\mathfrak{F}$). In particular, η_x and η_y are contained in K_o. Then by the UST Markov property (7.3), conditioned on K and the event $\mathscr{A}_\varepsilon^e$, the law of $\mathfrak{F}$ is equal to the union of K_o with a sample of the WUSF on $(G - K_c)/K_o$. In particular, by Kirchhoff's formula Theorem 7.5 we have that

$$\mathbf{P}(e \in \mathfrak{F} \mid \mathscr{A}_\varepsilon^e, K) \leq \mathcal{R}_{\text{eff}}^W(\eta_x \leftrightarrow \eta_y; G - K_c), \tag{7.10}$$

where in the last inequality we used the fact that gluing cannot increase the resistance (Corollary 2.30).

We will show that the last quantity tends to 0 as $\varepsilon \to 0$ which gives (7.8). To that aim, let v^x be the endpoint of the path η^x and let z_0 be the center of the $P(v_x)$. On the event $\mathscr{A}_\varepsilon^e$, for each $1 - |z_0| \leq r \leq 1/4$, we claim that the set $V_{z_0}(r, r)$, as defined in Definition 7.10, contains a path in G from η^x to η^y that is contained in $\mathcal{R} \cup \eta^x \cup \eta^y$ or an infinite simple path starting at η_x that is contained in $\mathcal{R} \cup \eta^x$. Either of these paths are therefore a path in $G - K_c$.

To see this, consider the arc $\mathsf{A}'(z_0, r) = \{z \in \overline{\mathbb{U}} : |z - z_0| = r\}$ viewed in the clockwise direction and let $\mathsf{A}(z_0, r)$ be the subarc beginning at the last intersection of $\mathsf{A}'(z_0, r)$ with a circle corresponding to a vertex in the trace of η^x, and ending at the first intersection after this time of $\mathsf{A}'(z_0, r)$ with either $\partial \mathbb{U}$ or a circle corresponding to a vertex in the trace of η^y (see Fig. 7.1). Hence, if $\mathscr{A}_\varepsilon^e$ holds, then the set of vertices of T whose circles in P intersect $\mathsf{A}(z_0, r)$ contains a path in T starting at η^x and ending η^y or does not end at all, for every $1 - |z_0| \leq r \leq 1/4$. To obtain a path in G rather than T we divert the path counterclockwise around each face of G. That is, whenever the path passes from a vertex u of G to a face f of G and then to a vertex v of G, we replace this section of the path with the list of vertices of G incident to f that are between u and v in the counterclockwise order. By Definition 7.10 this diverted path is in $V_{z_0}(r, r)$ and so this construction shows that the subgraph of $G - K_c$ induced by the set $V_{z_0}(r, r)$ contains a path from η^x to η^y or an infinite path from η^x, as claimed.

Let $r_i = C^{-i}$ for $i = 1, \ldots, N$ where $C < \infty$ the constant from Lemma 7.11 and $N = \lfloor \log_C(\varepsilon) \rfloor$. Assume without loss of generality that $C \geq 4$ so that $\varepsilon \leq r_i \leq 1/4$ for all $i = 1, \ldots, N$. By Lemma 7.11 the measures μ_{r_i} defined in Lemma 7.12 are supported on sets that are contained in the disjoint sets $V_z(r_i, 2r_i)$. Thus, by Lemma 7.9 and Lemma 7.12 we have

$$\mathcal{R}_{\text{eff}}^W\left(\eta^x \leftrightarrow \eta^y \cup \{\infty\}; G \setminus K_c\right) \leq \mathcal{E}\left(\frac{1}{N} \sum_{i=1}^N \mu_{r_i}\right) = \frac{1}{N^2} \sum_{i=1}^N \mathcal{E}(\mu_{r_i}) \leq \frac{B}{\log(1/\varepsilon)},$$

where $B < \infty$ is a constant depending only on the maximum degree. By symmetry we also have

$$\mathcal{R}_{\text{eff}}^W(\eta^y \leftrightarrow \eta^x \cup \{\infty\}; G - K_c) \leq \frac{B}{\log(1/\varepsilon)}.$$

Applying Lemma 7.8 and (7.10) gives

$$\mathbf{P}(e \in \mathfrak{F} \mid \mathcal{F}_K, \mathscr{A}_\varepsilon^e) \leq \frac{3B}{\log(1/\varepsilon)} \, .$$

We plug this estimate into (7.10) and take $\varepsilon \to 0$, which together with (7.9) shows that (7.8) holds, concluding our proof.   □

7.6 Exercises

1. Use Theorem 7.5 to show that on the 3-regular infinite tree $\mathbb{T}_3$ the probability measures $\mathsf{FUSF}_{\mathbb{T}_3}$ and the $\mathsf{WUSF}_{\mathbb{T}_3}$ are distinct.
2. Let $(G; \{r_e\})$ be a tree with edge resistances $\{r_e\}$ such that $\sum_{n \geq 1} r(e_n) = \infty$ for any simple infinite path $\{e_n\}_{n \geq 1}$ in G. Show that the free and wired uniform spanning forests coincide if and only if $(G; \{r_e\})$ is recurrent.
3. Let L_n be the ladder graph, that is, the vertex set is $\{1, \ldots, n\} \times \{a, b\}$ and the edges set is $\{[(i, a), (i, b)] : 1 \leq i \leq n\} \cup \{[(i, a), (i + 1, a)] : 1 \leq i \leq n - 1\} \cup \{[(i, b), (i + 1, b)] : 1 \leq i \leq n - 1\}$. Compute the limiting probability, as $n \to \infty$, that the edge $[(1, a), (1, b)]$ is in the UST of L_n.
4. Show that $\mathbb{Z}^3$ contains a transient subtree.

Chapter 8
Related Topics

In this chapter we briefly review some aspects of the literature on circle packing that unfortunately we do not have space to get into in depth in this course. We hope this will be useful as a guide to further reading.

1. **Double circle packing.** If one wishes to study planar graphs that are *not* triangulations, it is often convenient to work with *double circle packings*, which enjoy similar rigidity properties to usual circle packings, but for the larger class of **polyhedral** planar graphs. Here, a planar graph is polyhedral if it is both simple and **3-connected**, meaning that the removal of any two vertices cannot disconnect the graph. Double circle packings also satisfy a version of the ring lemma [45, Theorem 4.1], which means that they can be used to produce good straight-line embeddings of polyhedral planar graphs that have bounded face degrees but which are not necessarily triangulations.

   Let G be a planar graph with vertex set V and face set F. A double circle packing of G is a pair of circle packings $P = \{P_v : v \in V\}$ and $P^\dagger = \{P_f : f \in F\}$ satisfying the following conditions:

   (a) (*G* **is the tangency graph of** P.) For each pair of vertices u and v of G, the discs P_u and P_v are tangent if and only if u and v are adjacent in G.
   (b) ($G^\dagger$ **is the tangency graph of** $P^\dagger$.) For each pair of faces f and g of G, the discs P_f and P_g are tangent if and only if f and g are adjacent in $G^\dagger$.
   (c) (**Primal and dual circles are perpendicular**.) For each vertex v and face f of G, the discs P_f and P_v have non-empty intersection if and only if f is incident to v, and in this case the boundary circles of P_f and P_v intersect at right angles.

   See Fig. 8.1 for an illustration.

   Thurston's proof of the circle packing theorem also implies that every finite polyhedral planar graph admits a double circle packing. This was also shown by Brightwell and Scheinerman [13]. As with circle packings of triangulations, the double circle packing of any finite polyhedral planar map is unique up to

© The Author(s) 2020
A. Nachmias, *Planar Maps, Random Walks and Circle Packing*, Lecture Notes in Mathematics 2243, https://doi.org/10.1007/978-3-030-27968-4_8

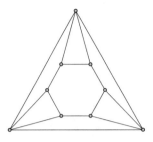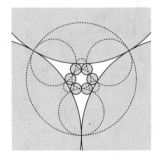

Fig. 8.1   A finite polyhedral plane graph (left) and its double circle packing (right). Primal circles are filled and have solid boundaries, dual circles have dashed boundaries

Möbius transformations or reflections. The theory of double circle packings in the infinite setting follows from the work of He [37], and is exactly analogous to the corresponding theory for triangulations. Indeed, essentially everything we have to say in these notes about circle packings of simple triangulations can be generalized to double circle packings of polyhedral planar maps (sometimes under the additional assumption that the faces are of bounded degree).

2. **Packing with other shapes**. A very powerful generalization of the circle packing theorem known as the *monster packing theorem* was proven by Schramm in his PhD thesis [76]. One consequence of this theorem is as follows: Let $T = (V, E)$ be a finite planar triangulation with a distinguished boundary vertex ∂. Specify a bounded, simply connected domain $D \subset \mathbb{C}$ with smooth boundary, and for each $v \in V \setminus \{\partial\}$ specify a strictly convex, bounded domain D_v with smooth boundary. Then there exists a collection of homotheties (compositions of translations and dilations) $\{h_v : v \in V\}$ such that

- If $u, v \in V \setminus \{\partial\}$ are distinct, then the closure of $h_v D_v$ and $h_u D_u$ have disjoint interiors, and intersect if and only if v and u are adjacent in T.
- If $v \in V \setminus \{\partial\}$, then the closure of $h_v D_v$ and $\mathbb{C} \setminus D$ have disjoint interiors, and intersect if and only if v is adjacent to ∂ in T.

In other words, we can represent the triangulation of T by a packing with arbitrary smooth convex shapes that are specified up to homothety (it is quite surprising at first that rotations are not needed). The full monster packing theorem also allows one to relax the smoothness and convexity assumptions above in various ways. The proof of the monster packing theorem is based upon Brouwer's fixed point theorem, and does not give an algorithm for computing the packing.

3. **Square tiling**. Another popular method of embedding planar graphs is the *square tiling*, in which vertices are represented by horizontal line segments and edges by squares; such square tilings can take place either in a rectangle, the plane, or a cylinder. Square tiling was introduced by Brooks et al. [14], and generalized to infinite planar graphs by Benjamini and Schramm [10]. Like circle packing,

Fig. 8.2 The square tiling of
the 7-regular triangulation

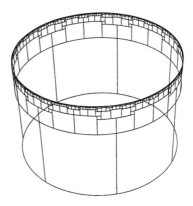

square tiling can be thought of as a discrete version of conformal mapping, and
in particular can be used to approximate the uniformizing map from a simply
connected domain with four marked boundary points to a rectangle. For studying
the random walk, a very nice feature of the square tiling that is not enjoyed by
the circle packing is that the height of a vertex in the cylinder is a harmonic
function, so that the height of a random walk is a martingale. Furthermore,
Georgakopoulos [28] observed that if one stops the random walk at the first time
it hits some height, then its horizontal coordinate at this time is uniform on the
circle (this takes some interpretation to make precise). Further works on square
tiling include [1, 28, 46] (Fig. 8.2).

Unlike circle packing, however, square tilings do not enjoy an analogue of
the ring lemma, and can be geometrically very degenerate. Indeed, it is possible
for edges to be represented by squares of zero area, and is also possible for two
distinct planar graphs to have the same square tiling. Furthermore, square tilings
are typically defined with reference to a specified root vertex, and it is difficult
to compare the two different square tilings of the same graph that are computed
with respect to different root vertices. These differences tend to mean that square
tilings are best suited to quite different problems than circle packing.

We also remark that a different sort of square tiling in which *vertices* are
represented by squares was introduced independently by Cannon et al. [15] and
Schramm [74].

4. **Multiply-connected triangulations.** Several works have studied generalizations
   of the circle packing theorem to triangulations that are either not simply
   connected or not planar. Most notably, He and Schramm [39] proved that every
   triangulation of a domain with countably many boundary components can be
   circle packed in a circle domain, that is, a domain all of whose boundary components are either circles or points: see Fig. 8.3 for examples. The corresponding
   statement for a triangulation of an *arbitrary* domain is a major open problem,
   and is closely related to the Koebe conjecture.

   Gurel-Gurevich, the current author, and Suoto [32] generalized the part of the
   He-Schramm Theorem concerning recurrence of the random walk as follows: A

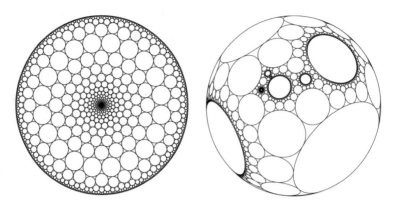

Fig. 8.3 Left: A circle packing in the multiply-connected circle domain $\mathbb{U} \setminus \{0\}$. Right: A circle packing in a circle domain with several boundary components

    bounded degree triangulation circle packed in a domain D is transient if and only if Brownian motion on D is transient, i.e. leaves D in finite time almost surely.

5. **Isoperimetry of planar graphs.** In [66], Miller, Teng, Thurston, and Vavasis used circle packing to give a new proof of the *Lipton-Tarjan planar separator theorem* [60], which concerns sparse cuts in planar graphs. Precisely, the theorem states that for any n-vertex planar graph, one can find a set of vertices of size at most $O(\sqrt{n})$ such that if this vertex set is deleted from the graph then every connected component that remains has size at most $3n/4$. More precisely, the authors of [66] showed that if one circle packs a planar graph in the unit sphere of $\mathbb{R}^3$, normalizes by applying an appropriate Möbius transformation, and takes a random plane passing through the origin in $\mathbb{R}^3$, then the set of vertices whose corresponding discs intersect the plane will have the desired properties with high probability.

    A related result of Jonasson and Schramm [47] concerns the *cover time* of planar graphs, i.e., the expected number of steps for a random walk on the graph to visit every vertex of the graph. They used circle packing to prove that the cover time of an n-vertex planar graph with maximum degree M is always at least $c_M n \log^2 n$ for some positive constant c_M depending only on M. This bound is attained (up to the constant) for large boxes $[-n, n]^2$ in $\mathbb{Z}^2$. In general, it is possible for n-vertex graphs to have cover time as small as $(1 + o(1))n \log n$.

6. **Boundary theory.** Benjamini and Schramm [9] proved that if P is a circle packing of a bounded degree triangulation in the unit disc $\mathbb{U}$, then the simple random walk on the circle packed triangulation converges to a point in the boundary of $\mathbb{U}$, and that the law of the limit point is non-atomic and has full support. (That is, the walk has probability zero of converging to any specific boundary point, and has positive probability of converging to any positive-length interval.) They used this result to deduce that a bounded degree planar graph admits non-constant bounded harmonic functions if and only if it is transient (equivalently, the invariant sigma-algebra of the random walk on the triangulation

is non-trivial if and only if the walk is transient), and in this case it also admits non-constant bounded harmonic functions of finite Dirichlet energy. They also gave an alternative proof of the same result using square tiling instead of circle packing in [10].

Indeed, given the result of Benjamini and Schramm, one may construct a non-constant bounded harmonic function h on T by taking any bounded, measurable function $f : \partial \mathbb{U} \to \mathbb{R}$ and defining h to be the *harmonic extension* of f, that is,

$$h(v) = \mathbf{E}_v \left[f \left(\lim_{n \to \infty} z(X_n) \right) \right],$$

where $\mathbf{E}_v$ denotes expectation taken with respect to the random walk X started at v, and $z(u)$ denotes the center of the circle in P corresponding to u. Angel et al. [6] proved that, in fact, *every* bounded harmonic function on a bounded degree triangulation can be represented in this way. In other words, the boundary $\partial \mathbb{U}$ can be identified with the **Poisson boundary** of the triangulation. Probabilistically, this means that the entire invariant σ-algebra of the random walk coincides with the σ-algebra generated by the limit point. They also proved the stronger result that $\partial \mathbb{U}$ can be identified with the **Martin boundary** of the triangulation. Roughly speaking, this means that every *positive* harmonic function on the triangulation admits a representation as the harmonic extension of some *measure* on $\partial \mathbb{U}$. A related representation theorem for harmonic functions of *finite Dirichlet energy* on bounded degree triangulations was established by Hutchcroft [43].

The results of [6] regarding the Poisson boundary followed earlier work by Georgakopoulos [28], which established a corresponding result for square tilings. Both results were revisited in the work of Hutchcroft and Peres [46], which gave a simplified and unified proof that works for both embeddings.

A parallel boundary theory for circle packings of **unimodular random triangulations** of *unbounded* degree was developed by Angel, Hutchcroft, the current author, and Ray in [7].

7. **Harnack inequalities, Poincaré inequalities, and comparison to Brownian motion.** The work of Angel et al. [6] also established various quite strong estimates for random walk on circle packings of bounded degree triangulations. Roughly speaking, these estimates show that the random walk behaves similarly to the image of a Brownian motion under a *quasi-conformal map*, that is, a bijective map that distorts angles by at most a bounded amount (i.e., maps infinitesimal circles to infinitesimal ellipses of bounded eccentricity). These estimates were central to their result concerning the Martin boundary of the triangulation, and are also interesting in their own right. Further related estimates have also been established by Chelkak [17].

Recent work of Murugan [67] has built further upon these methods to establish very precise control of the random walk on (graphical approximations of) various deterministic self-similar fractal surfaces.

8. **Liouville quantum gravity and the KPZ correspondence.** Statistical physics in two dimensions has been one of the hottest areas of probability theory in

recent years. The introduction of Schramm's SLE [75] and further breakthrough developments by Lawler, Schramm and Werner (see [53, 54] and the references within) on the one hand, and the application of discrete complex analysis, pioneered by Smirnov [79], on the other, have led to several breakthroughs and to the resolution of a number of long-standing conjectures. These include the conformally invariant scaling limits of critical percolation [77] and Ising models [78], and the determination of critical exponents and dimensions of sets associated with planar Brownian motion [53] (such as the frontier and the set of cut points). It is manifest that much progress will follow, possibly including the treatment of self-avoiding walk (the connective constant of the hexagonal lattice was calculated in the breakthrough work [22]), the $O(n)$ loop model and the Potts model. While the bulk of this body of work applies to specific lattices, there are many fascinating problems in extending results to arbitrary planar graphs.

The next natural step is to study the classical models of statistical physics in the context of random planar maps (see Le Gall's 2014 ICM proceedings [57]). There are deep conjectured connections between the behaviour of the models in the random setting versus the Euclidean setting, most significantly the KPZ formula of Knizhnik et al. [50] from conformal field theory. This formula relates the dimensions of certain sets in Euclidean geometry to the dimensions of corresponding sets in the random geometry. It may provide a systematic way to analyze models on the two dimensional Euclidean lattice: first study the model in the random geometry setting, where the Markovian properties of the underlying space make the model tractable; then use the KPZ formula to translate the critical exponents from the random setting to the Euclidean one.

Much of this picture is conjectural but a definite step towards this goal was taken in the influential paper of Duplantier and Sheffield [23]. Let us describe their formulation. Let G_n be a random triangulation on n vertices and consider its circle packing (or any other "natural" embedding) in the unit sphere. The embedding induces a random measure μ_n on the sphere by putting $\mu_n(A)$ to be the proportion of circle centers that are in A. The Duplantier-Sheffield conjecture asserts that the measures μ_n converge in distribution to a random measure μ on the sphere that has density given by an exponential of the Gaussian free field— the latter is carefully defined and constructed in [23]. This measure is what is known as *Liouville quantum gravity* (LQG).

Next, given a deterministic or random set K on the sphere, one can calculate its expected dimension using the random measure given by LQG, and using the usual Lebesgue measure—one gets two different numbers. Duplantier and Sheffield [23] obtain a quadratic formula allowing to compute one number from the other in the spirit of [50]; this is the first rigorous instance of the KPZ correspondence. It allows one to compute the dimension of random sets in the $\mathbb{Z}^2$ lattice (corresponding to Lebesgue measure) by first calculating the corresponding dimension in the random geometry setting and then appealing to the KPZ formula.

Many difficult models of statistical physics are tractable on a random planar map due to the inherent randomness of the space. For instance, it can be shown that the self avoiding walk on the UIPT behaves diffusively, that is, the endpoint of a self avoiding walk of length n is typically of distance $n^{1/2+o(1)}$ from the origin [19, 34]. A straightforward calculation with the KPZ formula allows one to predict that the typical displacement of the self-avoiding walk of length n on the lattice $\mathbb{Z}^2$ is $n^{3/4+o(1)}$—a notoriously hard open problem with endless simulations supporting it.

LQG and the KPZ correspondence thus pose a path to solving many difficult problems in classical two-dimensional statistical physics. We refer the interested reader to Garban's excellent survey [27] of the topic.

References

1. L. Addario-Berry, N. Leavitt, Random infinite squarings of rectangles. Ann. Inst. Henri Poincaré Probab. Stat. **52**(2), 596–611 (2016). MR3498002
2. D. Aldous, R. Lyons, Processes on unimodular random networks. Electron. J. Probab. **12**(54), 1454–1508 (2007). MR2354165
3. D. Aldous, J.M. Steele, The objective method: probabilistic combinatorial optimization and local weak convergence, in *Probability on Discrete Structures* (Springer, Berlin, 2004), pp. 1–72
4. O. Angel, Growth and percolation on the uniform infinite planar triangulation. Geom. Funct. Anal. **13**(5), 935–974 (2003). MR2024412
5. O. Angel, O. Schramm, Uniform infinite planar triangulations. Comm. Math. Phys. **241**(2–3), 191–213 (2003). MR2013797
6. O. Angel, M.T. Barlow, O. Gurel-Gurevich, A. Nachmias, Boundaries of planar graphs, via circle packings. Ann. Probab. **44**(3), 1956–1984 (2016). MR3502598
7. O. Angel, T. Hutchcroft, A. Nachmias, G. Ray, Unimodular hyperbolic triangulations: circle packing and random walk. Invent. Math. **206**(1), 229–268 (2016)
8. I. Benjamini, N. Curien, Simple random walk on the uniform infinite planar quadrangulation: subdiffusivity via pioneer points. Geom. Funct. Anal. **23**(2), 501–531 (2013). MR3053754
9. I. Benjamini, O. Schramm, Harmonic functions on planar and almost planar graphs and manifolds, via circle packings. Invent. Math. **126**(3), 565–587 (1996). MR1419007 (97k:31009)
10. I. Benjamini, O. Schramm, Random walks and harmonic functions on infinite planar graphs using square tilings. Ann. Probab. **24**(3), 1219–1238 (1996). MR1411492 (98d:60134)
11. I. Benjamini, O. Schramm, Recurrence of distributional limits of finite planar graphs. Electron. J. Probab. **6**(23), 1–13 (2001)
12. I. Benjamini, R. Lyons, Y. Peres, O. Schramm, Uniform spanning forests. Ann. Probab. **29**(1), 1–65 (2001). MR1825141 (2003a:60015)
13. G.R. Brightwell, E.R. Scheinerman, Representations of planar graphs. SIAM J. Discret. Math. **6**(2), 214–229 (1993). MR1215229 (95d:05043)
14. R.L. Brooks, C.A.B. Smith, A.H. Stone, W.T. Tutte, The dissection of rectangles into squares. Duke Math. J **7**, 312–340 (1940). MR0003040 (2,153d)
15. J.W. Cannon, W.J. Floyd, W.R. Parry, Squaring rectangles: the finite Riemann mapping theorem, in *The Mathematical Legacy of Wilhelm Magnus: Groups, Geometry and Special Functions (Brooklyn, NY, 1992)* (American Mathematical Society, Providence, 1994), pp. 133–212. MR1292901
16. P. Chassaing, G. Schaeffer, Random planar lattices and integrated superBrownian excursion. Probab. Theory Relat. Fields **128**(2), 161–212 (2004). MR2031225

© The Author(s) 2020

A. Nachmias, *Planar Maps, Random Walks and Circle Packing*, Lecture Notes in Mathematics 2243, https://doi.org/10.1007/978-3-030-27968-4

17. D. Chelkak, Robust discrete complex analysis: a toolbox. Ann. Probab. **44**(1), 628–683 (2016). MR3456348
18. R. Cori, B. Vauquelin, Planar maps are well labeled trees. Canad. J. Math. **33**(5), 1023–1042 (1981). MR638363
19. N. Curien, A. Caraceni, Self-avoiding walks on the UIPQ, Festschrift for Chuck Newman's 70th birthday (to appear)
20. R. Diestel, Graph theory, in *Fifth, Graduate Texts in Mathematics*, vol. 173 (Springer, Berlin, 2017). MR3644391
21. P. Doyle, Z.-X. He, B. Rodin, Second derivatives of circle packings and conformal mappings. Discrete Comput. Geom. **11**(1), 35–49 (1994). MR1244888
22. H. Duminil-Copin, S. Smirnov, The connective constant of the honeycomb lattice equals, $\sqrt{2 + \sqrt{2}}$. Ann. Math. (2) **175**(3), 1653–1665 (2012). MR2912714
23. B. Duplantier, S. Sheffield, Liouville quantum gravity and KPZ. Invent. Math. **185**(2), 333–393 (2011). MR2819163
24. R. Durrett, Probability: theory and examples, in *Fourth, Cambridge Series in Statistical and Proba-bilistic Mathematics*, vol. 31 (Cambridge University Press, Cambridge, 2010). MR2722836
25. I. Fáry, On straight line representation of planar graphs. Acta Univ. Szeged. Sect. Sci. Math. **11**, 229–233 (1948). MR0026311
26. Z. Gao, L.B. Richmond, Root vertex valency distributions of rooted maps and rooted triangulations. Eur. J. Comb. **15**(5), 483–490 (1994). MR1292958
27. C. Garban, Quantum gravity and the KPZ formula [after Duplantier-Sheffield]. Astérisque **352** (2013), Exp. No. 1052, ix, 315–354. Séminaire Bourbaki, vol. 2011/2012. Exposés 1043–1058. MR3087350
28. A. Georgakopoulos, The boundary of a square tiling of a graph coincides with the Poisson boundary. Invent. Math. **203**(3), 773–821 (2016). MR3461366
29. J.T. Gill, Doubling metric spaces are characterized by a lemma of Benjamini and Schramm. Proc. Am. Math. Soc. **142**(12), 4291–4295 (2014). MR3266996
30. J.T. Gill, S. Rohde, On the Riemann surface type of random planar maps. Rev. Mat. Iberoam. **29**(3), 1071–1090 (2013). MR3090146
31. O. Gurel-Gurevich, A. Nachmias, Recurrence of planar graph limits. Ann. Math. (2) **177**(2), 761–781 (2013). MR3010812
32. O. Gurel-Gurevich, A. Nachmias, J. Souto, Recurrence of multiply-ended planar triangulations. Electron. Commun. Probab. **22**(5), 6 (2017). MR3607800
33. E. Gwynne, J. Miller, Random walk on random planar maps: spectral dimension, resistance, and displacement, Preprint (2017)
34. E. Gwynne, J. Miller, Convergence of the self-avoiding walk on random quadrangulations to $sle_8/3$ on $\sqrt{8/3}$-liouville quantum gravity.
35. E. Gwynne, J. Miller, S. Sheffield, The Tutte embedding of the mated-crt map converges to liouville quantum gravity, Preprint (2017)
36. Z.-X. He, An estimate for hexagonal circle packings. J. Differ. Geom. **33**(2), 395–412 (1991). MR1094463
37. Z.-X. He, Rigidity of infinite disk patterns. Ann. Math. **149**, 1–33 (1999)
38. Z.-X. He, B. Rodin, Convergence of circle packings of finite valence to Riemann mappings. Comm. Anal. Geom. **1**(1), 31–41 (1993). MR1230272
39. Z.-X. He, O. Schramm, Fixed points, Koebe uniformization and circle packings. Ann. Math. (2) **137**(2), 369–406 (1993). MR1207210 (96b:30015)
40. Z.-X. He, O. Schramm, Hyperbolic and parabolic packings. Discret. Comput. Geom. **14**(2), 123–149 (1995). MR1331923 (96h:52017)
41. Z.-X. He, O. Schramm, On the convergence of circle packings to the Riemann map. Invent. Math. **125**(2), 285–305 (1996). MR1395721
42. Z.-X. He, O. Schramm, The C^∞-convergence of hexagonal disk packings to the Riemann map. Acta Math. **180**(2), 219–245 (1998). MR1638772

43. T. Hutchcroft, Harmonic dirichlet functions on planar graphs (2017). arXiv preprint. arXiv:1707.07751
44. T. Hutchcroft, A. Nachmias, Uniform spanning forests of planar graphs. Forum Math. Sigma 7, e29 (2019)
45. T. Hutchcroft, A. Nachmias, Uniform spanning forests of planar graphs (2016). http://arxiv.org/abs/1603.07320
46. T. Hutchcroft, Y. Peres, Boundaries of planar graphs: a unified approach (2015). arXiv preprint arXiv:1508.03923
47. J. Jonasson, O. Schramm, On the cover time of planar graphs. Electron. Comm. Probab. **5**, 85–90 (2000) (Electronic). MR1781842 (2001g:60170)
48. J.A. Kelner, J.R. Lee, G.N. Price, S.-H. Teng, Metric uniformization and spectral bounds for graphs. Geom. Funct. Anal. **21**(5), 1117–1143 (2011). MR2846385
49. G. Kirchhoff, Ueber die auflösung der gleichungen, auf welche man bei der untersuchung der linearen vertheilung galvanischer ströome gefuührt wird. Annalen der Physik **148**(12), 497–508 (1847)
50. V.G. Knizhnik, A.M. Polyakov, A.B. Zamolodchikov, Fractal structure of 2D-quantum gravity. Modern Phys. Lett. A **3**(8), 819–826 (1988). MR947880
51. P. Koebe, Kontaktprobleme der konformen abbildung (Hirzel, German, 1936)
52. M. Krikun, Local structure of random quadrangulations, Preprint (2005)
53. G.F. Lawler, O. Schramm, W. Werner, Values of Brownian intersection exponents. I. Half-plane exponents. Acta Math. **187**(2), 237–273 (2001). MR1879850
54. G.F. Lawler, O. Schramm, W. Werner, Conformal invariance of planar loop-erased ran- dom walks and uniform spanning trees. Ann. Probab. **32**(1B), 939–995 (2004). MR2044671
55. J.-F. Le Gall, The topological structure of scaling limits of large planar maps. Invent. Math. **169**(3), 621–670 (2007). MR2336042
56. J.-F. Le Gall, Uniqueness and universality of the Brownian map. Ann. Probab. **41**(4), 2880–2960 (2013). MR3112934
57. J.-F. Le Gall, Random geometry on the sphere, in *Proceedings of the International Congress of Mathematicians—Seoul 2014*, vol. 1 (2014), pp. 421–442. MR3728478
58. J.-F. Le Gall, G. Miermont, Scaling limits of random trees and planar maps, in *Probability and Statistical Physics in Two and More Dimensions* (2012), pp. 155–211. MR3025391
59. J. Lee, Conformal growth rates and spectral geometry on distributional limits of graphs, Preprint (2017)
60. R.J. Lipton, R.E. Tarjan, Applications of a planar separator theorem. SIAM J. Comput. **9**(3), 615–627 (1980). MR584516 (82e:68067)
61. R. Lyons, Y. Peres, Probability on trees and networks, in *Cambridge Series in Statistical and Probabilistic Mathematics*, vol. 42 (Cambridge University Press, New York, 2016). MR3616205
62. J.-F. Marckert, A. Mokkadem, Limit of normalized quadrangulations: the Brownian map. Ann. Probab. **34**(6), 2144–2202 (2006). MR2294979
63. A. Marden, B. Rodin, On thurston's formulation and proof of andreev's theorem. in *Computational Methods and Function Theory* (1990), pp. 103–115
64. G.H. Meisters, Polygons have ears. Am. Math. Monthly **82**, 648–651 (1975). MR0367792
65. G. Miermont, The Brownian map is the scaling limit of uniform random plane quadrangulations. Acta Math. **210**(2), 319–401 (2013). MR3070569
66. G.L. Miller, S.-H. Teng, W. Thurston, S.A. Vavasis, Separators for sphere-packings and nearest neighbor graphs. J. ACM **44**(1), 1–29 (1997)
67. M. Murugan, Quasisymmetric uniformization and heat kernel estimates (2018). arXiv preprint arXiv:1803.11296
68. R. Pemantle, Choosing a spanning tree for the integer lattice uniformly. Ann. Probab. **19**(4), 1559–1574 (1991). MR1127715 (92g:60014)
69. Y. Peres, Probability on trees: an introductory climb, in *Lectures on Probability Theory and Statistics (Saint-Flour, 1997)* (Springer, Berlin, 1999), pp. 193–280. MR1746302

70. B. Rodin, D. Sullivan, The convergence of circle packings to the Riemann mapping. J. Differ. Geom. **26**(2), 349–360 (1987). MR906396 (90c:30007)

71. S. Rohde, Oded Schramm: from circle packing to SLE. Ann. Probab. **39**(5), 1621–1667 (2011). MR2884870

72. G. Schaeffer, Conjugaison darbres et cartes combinatoires alatoires, Ph.D. Thesis, Universit Bordeaux I (1998)

73. O. Schramm, Rigidity of infinite (circle) packings. J. Am. Math. Soc. **4**(1), 127–149 (1991). MR1076089 (91k:52027)

74. O. Schramm, Square tilings with prescribed combinatorics. Israel J. Math. **84**(1–2), 97–118 (1993). MR1244661

75. O. Schramm, Scaling limits of loop-erased random walks and uniform spanning trees. Israel J. Math. **118**, 221–288 (2000). MR1776084

76. O. Schramm, Combinatorically prescribed packings and applications to conformal and quasi-conformal maps (2007). arXiv preprint arXiv:0709.0710

77. S. Smirnov, Critical percolation in the plane: conformal invariance, Cardy's formula, scaling limits. C. R. Acad. Sci. Paris Sér. I Math. **333**(3), 239–244 (2001). MR1851632

78. S. Smirnov, Conformal invariance in random cluster models. I. Holomorphic fermions in the Ising model. Ann. Math. (2) **172**(2), 1435–1467 (2010). MR2680496

79. S. Smirnov, Discrete complex analysis and probability, in *Proceedings of the International Congress of Mathematicians*, vol. I (2010), pp. 595–621. MR2827906

80. K. Stephenson, A probabilistic proof of Thurston's conjecture on circle packings. Rend. Sem. Mat. Fis. Milano **66**(1996), 201–291 (1998). MR1639851

81. K. Stephenson, *Introduction to Circle Packing* (Cambridge University Press, Cambridge, 2005). The theory of discrete analytic functions. MR2131318 (2006a:52022)

82. W.P. Thurston, The geometry and topology of 3-manifolds, in *Princeton Lecture Notes.* (Unknown Month 1978).

83. W.T. Tutte, A census of planar triangulations. Canad. J. Math. **14**, 21–38 (1962). MR0130841

Index

© The Author(s) 2020
A. Nachmias, *Planar Maps, Random Walks and Circle Packing*, Lecture Notes in Mathematics 2243, https://doi.org/10.1007/978-3-030-27968-4

LECTURE NOTES IN MATHEMATICS  Springer

Editors in Chief: J.-M. Morel, B. Teissier;

Editorial Policy

1. Lecture Notes aim to report new developments in all areas of mathematics and their applications – quickly, informally and at a high level. Mathematical texts analysing new developments in modelling and numerical simulation are welcome.

   Manuscripts should be reasonably self-contained and rounded off. Thus they may, and often will, present not only results of the author but also related work by other people. They may be based on specialised lecture courses. Furthermore, the manuscripts should provide sufficient motivation, examples and applications. This clearly distinguishes Lecture Notes from journal articles or technical reports which normally are very concise. Articles intended for a journal but too long to be accepted by most journals, usually do not have this "lecture notes" character. For similar reasons it is unusual for doctoral theses to be accepted for the Lecture Notes series, though habilitation theses may be appropriate.

2. Besides monographs, multi-author manuscripts resulting from SUMMER SCHOOLS or similar INTENSIVE COURSES are welcome, provided their objective was held to present an active mathematical topic to an audience at the beginning or intermediate graduate level (a list of participants should be provided).

   The resulting manuscript should not be just a collection of course notes, but should require advance planning and coordination among the main lecturers. The subject matter should dictate the structure of the book. This structure should be motivated and explained in a scientific introduction, and the notation, references, index and formulation of results should be, if possible, unified by the editors. Each contribution should have an abstract and an introduction referring to the other contributions. In other words, more preparatory work must go into a multi-authored volume than simply assembling a disparate collection of papers, communicated at the event.

3. Manuscripts should be submitted either online at www.editorialmanager.com/lnm to Springer's mathematics editorial in Heidelberg, or electronically to one of the series editors. Authors should be aware that incomplete or insufficiently close-to-final manuscripts almost always result in longer refereeing times and nevertheless unclear referees' recommendations, making further refereeing of a final draft necessary. The strict minimum amount of material that will be considered should include a detailed outline describing the planned contents of each chapter, a bibliography and several sample chapters. Parallel submission of a manuscript to another publisher while under consideration for LNM is not acceptable and can lead to rejection.

4. In general, **monographs** will be sent out to at least 2 external referees for evaluation.

   A final decision to publish can be made only on the basis of the complete manuscript, however a refereeing process leading to a preliminary decision can be based on a pre-final or incomplete manuscript.

   Volume Editors of **multi-author works** are expected to arrange for the refereeing, to the usual scientific standards, of the individual contributions. If the resulting reports can be forwarded to the LNM Editorial Board, this is very helpful. If no reports are forwarded or if other questions remain unclear in respect of homogeneity etc, the series editors may wish to consult external referees for an overall evaluation of the volume.

5. Manuscripts should in general be submitted in English. Final manuscripts should contain at least 100 pages of mathematical text and should always include

   - a table of contents;
   - an informative introduction, with adequate motivation and perhaps some historical remarks: it should be accessible to a reader not intimately familiar with the topic treated;
   - a subject index: as a rule this is genuinely helpful for the reader.
   - For evaluation purposes, manuscripts should be submitted as pdf files.

6. Careful preparation of the manuscripts will help keep production time short besides ensuring satisfactory appearance of the finished book in print and online. After acceptance of the manuscript authors will be asked to prepare the final LaTeX source files (see LaTeX templates online: https://www.springer.com/gb/authors-editors/book-authors-editors/manuscriptpreparation/5636) plus the corresponding pdf- or zipped ps-file. The LaTeX source files are essential for producing the full-text online version of the book, see http://link.springer.com/bookseries/304 for the existing online volumes of LNM). The technical production of a Lecture Notes volume takes approximately 12 weeks. Additional instructions, if necessary, are available on request from lnm@springer.com.

7. Authors receive a total of 30 free copies of their volume and free access to their book on SpringerLink, but no royalties. They are entitled to a discount of 33.3 % on the price of Springer books purchased for their personal use, if ordering directly from Springer.

8. Commitment to publish is made by a *Publishing Agreement*; contributing authors of multiauthor books are requested to sign a *Consent to Publish form*. Springer-Verlag registers the copyright for each volume. Authors are free to reuse material contained in their LNM volumes in later publications: a brief written (or e-mail) request for formal permission is sufficient.

Addresses:
Professor Jean-Michel Morel, CMLA, École Normale Supérieure de Cachan, France
E-mail: moreljeanmichel@gmail.com

Professor Bernard Teissier, Equipe Géométrie et Dynamique,
Institut de Mathématiques de Jussieu – Paris Rive Gauche, Paris, France
E-mail: bernard.teissier@imj-prg.fr

Springer: Ute McCrory, Mathematics, Heidelberg, Germany,
E-mail: lnm@springer.com

Printed in the United States
By Bookmasters